**Billion-Dollar Fish**

# BILLION-DOLLAR FISH

## THE UNTOLD STORY OF ALASKA POLLOCK

~~~~~~~~~~~~~ Kevin M. Bailey ~~~~~~~~~~~~~

THE UNIVERSITY OF CHICAGO PRESS   CHICAGO AND LONDON

**Kevin M. Bailey** is the founding director of Man & Sea Institute and affiliate professor at the University of Washington. He formerly was a senior scientist at the Alaska Fisheries Science Center. He has published over 100 papers, mostly about Alaska pollock.

The University of Chicago Press, Chicago 60637
The University of Chicago Press, Ltd., London
© 2013 by The University of Chicago
All rights reserved. Published 2013.
Printed in the United States of America

22  21  20  19  18  17  16  15  14  13      1  2  3  4  5

ISBN-13: 978-0-226-02234-5 (cloth)
ISBN-13: 978-0-226-02248-2 (e-book)

Library of Congress Cataloging-in-Publication Data

Bailey, Kevin McLean, author.
    Billion-dollar fish : the untold story of Alaska pollock / Kevin M. Bailey.
        pages cm
    Includes bibliographical references and index.
    ISBN 978-0-226-02234-5 (cloth : alkaline paper) — ISBN 978-0-226-02248-2 (e-book)
1. Pollock fisheries—History—20th century.   2. Walleye pollock—Effect of fishing on.   I. Title.
    SH351.W32B35 2013
    639.3'772—dc23
                                                                            2012044795

♾ This paper meets the requirements of ANSI/NISO
Z39.48-1992 (Permanence of Paper).

# Contents

Illustrations follow pages 58 and 138.

# Preface

The Alaska pollock fishery developed with explosive force, rising from obscurity about forty years ago to become the world's largest food fishery within a decade. The fishermen's invasion of the arctic seas to catch the fish with snow-white meat resembled a gold rush. At the time, knowledge of the resource lagged far behind the process of exploitation, depletion, and discovery of new stocks. Scientists could not keep up with an industry on steroids. By the late 1980s, ships to fish for pollock were being built at a furious pace. Empires were created and fortunes were made and lost. In some regions, stocks were overfished and depleted.

Many people witnessed the pollock story from the beginning, but few contributed to making the history. I was a biologist trying to understand changes in the population, not a fisherman, politician, or manager, all of whom influence the course of events. Looking back, I realize that as an observer on the sidelines, I had stumbled into the shadow of the world's largest food fishery and along the way witnessed one of the largest fishery collapses in history, that of the "Donut Hole" stock of pollock. Since my own career spans much of the development of the fishery, I became motivated to discover what happened during the time I was busily engaged in science.

My intent in telling the story of Alaska pollock is to describe not only what happened, but also why things happened. I learned that both opportunity and personalities drive the story of the fishery, give it life, and make it interesting. Without exception, I appreciate and respect the many people I talked with during the research for this book. I learned something from each. Many of the people involved in the early years of the US fishery are still alive, and I was fortunate to interview some of them.

When riches are involved, people battle over them. In the case of pollock, there were fights over many issues related to the resource. As the path of this book developed, the conflicts among groups became tangible. I tried to get opinions from each side of the issues and to present them as fairly as I could. At times I felt like a hapless observer who had walked into the middle of a duel between two offended parties. It's hard to figure out who is right. Everyone is the hero of his own side of the story.

I tell the story of the pollock fishery in a narrative style, braiding interviews and stories with descriptions of science, policy, and events. The story starts with a prologue, which is a description of the circuitous route I took in my career in fisheries biology that acquainted me with pollock. The Introduction (Chapter 1) summarizes the main subjects and themes that follow. Chapter 2 sets the stage of the global and historical scene of fishing, the broader landscape in which the pollock story played out.

The fishery was largely developed by Japan, as described in Chapter 3. The direct role of the Supreme Commander of Allied Powers in the developing fishery was surprising. A growing awareness of the richness of the fisheries in the Bering Sea and the declaration of a 200-mile fisheries zone in 1976 leads into the text on Americanization (Chapter 4).

The Americanization process squeezed the enormous fishing capacity of Japan, Korea, and the Union of Soviet Socialist Republics (USSR) into the international high seas. Chapter 5 documents its demise in the "Donut Hole" due to overfishing. Chapter 6 tells about the massive investment of Norwegian banks in the fishery and the critical role of Norwegian immigrants. Then there is a short break in the narrative to catch up on what we've learned about pollock that we didn't know previously (Chapter 7) and the changing global landscape of fisheries (Chapter 8).

Americanization of the fishery brought several new problems. Chapter 9 describes the struggle between industry and the environmental movement. Chapter 10 describes the strife within the industry caused by a race that involved too many boats catching a limited number of fish.

After the American Fisheries Act passed in 1998, the waters of the pollock industry calmed. But new issues keep surfacing. Chapter 11 describes the struggle for fairness in setting up private quotas, resources given to western Alaskan coastal communities, and issues of salmon bycatch, management councils, and ecosystem impacts. Finally, Chapter 12 paints a scene of the future.

Over the course of this project over fifty people were formally interviewed or engaged in long discussions. In most cases, their participation also involved a review of the notes I had taken and often follow-up correspondence, phone calls, or further discussions. I thank the following people for allowing themselves to be interviewed and for their openness: Tim Thomas, Mick Stevens, Becca Robbins-Gisclair, Fred Munson, Vera Schwach, Wally Pereyra, Marty Nelson, Ken Stump, John Warrenchuk, Rune Hornnes, Tony Allison, Lee Alverson, Linda Behnken, Dave Fluharty, Dan Huppert, Chris McReynolds, John Sjong, Bernt Bodal, Brent Paine, Dorothy Lowman, Magne Nes, Mike Weber, John Gruver, Rod Fujita, Tor Tollessen, Paul MacGregor, Jan Jacobs, Susanne Iudicello, Tim Smith, Zoya Johnson, Chris Mackie, Ed Miles, Lowell Fritz, Jena Carter, Larry Merculieff, Boris Olich, Howard Carlough, Jessie Gharrett, Doug Dixon, Leif Mannes, Artur Dacruz, Kaare Ness, Mary Furuness, Becky Mansfield, Dave Fraser, Mike Zimny, Ed Luttrell, Jim Ianelli, Alan Longhurst, Joe Plesha, Marc Wells, Vera Agostini, Pat Shanahan, Erik Breivik, John Bundy, and several anonymous fishermen. I attempted to contact or interview many others and was not successful.

I am grateful for the repeated discussions, suggestions, and/or encouragement of Monica Orellana, Mike Canino, Wayne Palsson, Jeff Napp, Carmel Finley, Layne Maheu, Gary Stauffer, Lorenzo Ciannelli, Mary Hunsicker, Bob Francis, Daniel Sloan, Mike Macy, Suam Kim, Anne Hollowed, neighbors, and others. I also appreciated Nick O'Connell's Writer's Workshop and the support of my classmates. The following people relayed information/contacts or gave comments: Stan Senner, Dorothy Childers, Paul Dye, Bill Eichbaum, Gary Stauffer, Mike Macy, Doug Dixon, Carmel Finley, Gordy Swartzman (for the Dickens quote), Akira Nishimura, Katie Flynn-Jambeck, Chris Wilson (for suggesting that I write a book on pollock), and Dave King.

Photos were provided by Jan Jacobs of American Seafoods, Aaron Barnett of Golden Alaska Seafoods, Linda Lowry and Elizabeth Masoni of Unalaska, Annette Dougherty, Larry Murphy of Zendog Studios, and Michelle Ruge and Patricia Nelson of the Fisheries Management and Analysis (FMA) Division of the Alaska Fisheries Science Center. The sketch at the end of the book was created by Mattias Bailey. John Sabella gave permission to use interviews from "Centuries of Fish" and Tony Allison gave me access to his collection of interviews with Clem Tillion and Lee Alverson. Translation of Norwegian books was provided by Heather Ione-Short of the Scandinavian Studies Department, University of Washington.

I appreciate editorial criticism and advice from Barbara Sjoholm. Jeff Napp, Jeff Buckel, Tim Smith, Gary Duker, and Nate Bacheler read the complete manuscript and made many helpful comments and corrections. Michele Rudnick, Dan Sloan, Pat Chaney, Janet Duffy-Anderson, Layne Maheu, Matt Wilson, Mike Macy, Jan Hartung, Wayne Palsson, Mary Hunsicker, Lorenzo Ciannelli, Knut Vollset, and Lowell Fritz read certain sections and provided suggestions and encouragement.

I had the support of my family, who patiently listened to my stories repeated many times over or even read them. These include Andres Hermosilla, Elizabeth Corman, Monica Orellana, Renato Orellana, Jan Hartung, Spencer Eric Bailey, and Mattias Bailey.

I also benefited from the support of my colleagues at the Alaska Fisheries Science Center, especially in the Fisheries Oceanography and Coordinated Investigations Program, headed by Art Kendall and then by Jeff Napp. I have been lucky to have many mentors during my career, including Jean Dunn, Donald McKernan, Frieda Taub, Warren Wooster, John Blaxter, Bob Francis, and Gary Stauffer.

I was inspired by Tim Smith's book *Scaling Fisheries*, Carmel Finley's *All the Fish in the Sea*, Dayton Alverson's *Race to the Sea*, and Alan Longhurst's *Mismanagement of Marine Fisheries*. Each of these touched on some aspect of the story I wanted to tell.

Finally, Christie Henry and the publication team (Mary Corrado, Micah Fehrenbacher, Abby Collier, and Matt Avery) at the University of Chicago Press gave me the opportunity to write the book and provided encouragement, editing, and advice.

# Prologue: Fishing Lessons

*Whether I shall turn out to be the hero of my own life, or whether that station will be held by anybody else, these pages must show.*

CHARLES DICKENS, in *David Copperfield*

Each of us is the protagonist of our own story, a tale that is usually set in the context of a much larger scene. My own narrative is like that, a small ripple riding on a large wave. The story of how my ripple was swallowed by the wave of the Alaska pollock fishery is as follows.

One morning I woke from a deep sleep, confused and disoriented. I lay in a strange bed that was too short, while the room around me pitched up and down. Rubbing my eyes, I thought, "Where am I? How did I get here?" As the fog lifted from my mind, I remembered I was on a Japanese fishing boat in the middle of the Bering Sea.

I felt a little queasy as I weaved my way across the rising and falling floor toward the door. I opened it and headed for some fresh air on deck. In the corridor outside there was a body on a gurney. The ship's doctor smiled at me as he rolled the body into the hospital across the hall. "Ohayo gozaimasu." I recognized him from dinner the night before. He was stocky, with wire-rimmed glasses, a thick graying bottle-brush crew-cut, and teeth inlaid with glinting gold. Something about him made me wonder if he was a real doctor. I thought, "Why would a *real doctor* be condemned to a ship in the middle of the Bering Sea?" I entertained the thought that maybe he was a large-animal veterinarian being punished for committing some unspeakable offense. Did he accidentally castrate the emperor's prize stallion?

My immediate concern was that the doctor was removing everyone's appendix on the mother ship and her fleet of catcher vessels.

The body on the gurney was his most recent victim. I'd read that they do this surgery preemptively on astronauts to short-circuit emergencies, but I was hoping to hang onto all of my own organs—especially given what he'd done to the emperor's horse. I was nervous to complain about even minor maladies, like say an ingrown toe nail. I imagined the doctor saying in a thick accent, "Ohhh, yes, Mr. Bailey-san," and then with a glint of anticipation in his eye, "We need to remove appendix, please."

Thus began my apprenticeship as a fish biologist in 1974. I'd been plucked from the beach in California and plopped down on a Japanese floating factory in a cold and stormy arctic sea. It could not have been a more abrupt and unexpected change for me. I had lived my whole life near the coast in central California. I went to college in Santa Barbara. Gone was the warm sand, the fragrance of flowers and suntan lotion, the sound of waves crashing on the beach. In their place was bitter cold, fishy stench, diesel fumes, and the noise of machinery, cranes, and seagulls.

Just weeks before, I hardly knew there was a career option in fisheries biology. One day I was visiting my sister in Seattle. It seems like the next day I was in the Bering Sea. By chance I had walked by the "Northwest and Alaska Fisheries Center." With a degree in biology, I was unemployed and had no prospects. I thought, "Why not?" and strolled inside on a whim. I approached the first person I saw and asked if they employed biologists. He sized me up with a quick look, and said, "As a matter of fact, we do. I have a job open right now." He interviewed me on the spot. The process was suspiciously brief, and he hired me to go out on a fishing boat in the North Pacific Ocean for a month. The trip ended up lasting almost four months.

The US government started putting American scientists on Japanese fishing vessels in 1973 as part of a treaty to monitor Japanese groundfish and crab catches in the Bering Sea. I was the third person hired specially as a scientist observer of the foreign fleet. They planned to station me on a mother ship called the *Keiko Maru*, which serviced a fleet of smaller catcher vessels fishing for tanner crab. My job was to monitor the species composition of the catch and take some biological measurements.

The scientists were sent out in pairs, and I was partnered with an older biologist from the National Marine Fisheries Service. Ken planned to spend two weeks with me on the *Keiko Maru*, by which time I would be fully trained in the tools of the trade. Then another inexperienced scientist would replace Ken, I would train him, and so on.

Nowadays government officials are briefed about receiving gifts and favors. Not so in my case. We flew to Tokyo and spent two weeks there waiting for "transport" to the Bering Sea, during which time we were entertained by representatives of the Japanese government and of the corporation that owned my ship. We were given gifts of pearls and cloisonné, fed Kobe beef at expensive restaurants, and taken to geisha houses and "companion" bars. (I suppose at this stage whatever political aspirations I may have had are over.)

I didn't realize the cultural implications of the offerings at the time. I have since learned about the tradition of gift giving in Japan, a practice called *omiyage*. The exchange of presents implies a social obligation, a trading of favors, and a bond of allegiance that dates back to the time of shoguns and samurais.

We left Tokyo on a small transport ship that carried mail, personnel, and supplies out to the fleet in the Bering Sea. This little boat was built before World War II when the average height of Japanese men was 5′4." I had to duck going through passageways. The toilet was not much more than a hole in the deck with two handles strategically placed on the wall in front. Its use was tricky in a rolling sea, requiring a basic knowledge of physics and precise coordination. The surroundings told the story of several passengers who hadn't mastered those skills.

Finally we arrived at our ship in the Bering Sea. It was late winter, cold, gray, and stormy, as is normal there. Crabs were caught in pots, and the pots were emptied into bags of netting on small catcher boats. The catcher boats would accumulate about twenty bags, each weighing as much as 500 kg, and then deliver them to the mother ship where they were processed, mostly into frozen crab legs.

For the first weeks, we worked, ate, and slept uneventfully. The ship's officers, her crew, or the factory management entertained us almost every night. These evenings were usually small festive parties with drinks and special foods. Ken often sang sea shanties at these events. I can still remember his voice, which strayed off-course as it progressed up the scale. Ken stretched his neck and cocked his head as he cawed out the high notes. Having foreigners on the boat must have been a form of entertainment to the crew. We were popular everywhere we went on the ship. The crew greeted us and tried out their English. "You have girlfriend?" "You like Japanese girl?" "You like Japanese whiskey?" Life onboard was good.

When Ken left, a new scientist named Paul replaced him. Paul harbored suspicions about the Japanese fishing off our coast and had a different agenda for our role on the ship. He had the mindset of an accountant and wanted accurate numbers. He continuously jotted his observations in a notepad and took pictures. Paul began to notice irregularities in the ship's catching and reporting system. Once he pointed out the problems, I noticed things I hadn't seen before.

We planned to confirm the catch tonnage reported to us by the ship's management instead of accepting them on faith. To do this, we counted all of the crabs in several bags delivered to the mother ship by the catcher boats—a number we multiplied by the average weight of individual crabs and the number of bags delivered to obtain the total tonnage landed. We discovered that our estimates were about double the catches reported to us. All signs pointed to tampering of the scales.

Then we started secretly keeping count of deliveries. We found that the number of bags arriving onboard often conflicted with the number given to us by the ship's personnel. The reported numbers were uniformly underestimated.

Although our role on the ship was "to observe," we were offended by the obvious cheating. We confronted the ship's fishing manager with the discrepancies. He got angry and hostile.

Overnight, our friendly relations with the fishermen soured. The crew of about 300 men shunned us and a tense situation developed over the next couple weeks. There were no more invitations to parties. Friendly greetings were replaced by sneers, angry voices, and even threatening gestures.

Our work on deck turned chaotic. The catcher boats now delivered the bags of crab at different points all over the mother ship and at all hours. Previously they had delivered to one main site on the deck and only during daylight hours. I was shocked and disillusioned that my fishermen "tomodachi" would be so dishonest. How could my friends of the last months become part of an unfriendly mob?

The ship's management directed most of their aggression at Paul. He would be separately dragged into meetings and harangued in Japanese, which he didn't understand. During these keel-haulings when I was alone on deck, the deliveries from the catcher vessels intensified. Our direct and confrontational approach came across as an accusation of cheating and dishonor. We had not been subtle. We were no longer a novelty but an annoyance.

Under the harsh conditions of isolation and animosity, and fearing for his life, Paul gave me a note intended for his parents in case he disappeared. The waters of the Bering Sea in winter are among the roughest in the world. It can get crazy out there. Sea spray freezes on the metal. Gusts of wind lift you off the deck. It wasn't hard to imagine being blown like a floating leaf into the sea. In the Bering, it isn't uncommon for sailors to disappear overboard.

I remember thinking that Paul was tense and on edge, maybe more so than me. The ship didn't feel safe anymore. The strain was compounded by limited communication with our office in Seattle, which was by telegraph. Since we sent messages through the ship's radioman, we could never be sure they were transmitted. We felt isolated and far from help. The adventure had turned into a nightmare.

After another week of tension, Paul couldn't take it anymore and arranged to leave for home. I received a cable from headquarters asking me to spend another month onboard in order to receive and train our replacements. I felt that I didn't have much choice. Paul boarded a small ship that tied up alongside us and sailed off; I never heard from him again. Collectively the ship exhaled a breath of relief when he left.

Not long after Paul departed, and now alone, I had a minor medical emergency that broke the logjam of tension still lingering on the ship. I became aware that I was constipated. In the day-to-day routine aboard ship, I lost track of just how long this had been going on, or maybe I should say, had not been going on. After a few more days of waiting, I paid a nervous visit to the doctor. Once again he brought up the issue of my troublesome appendix, and again I declined his offer to eliminate my pesky organ forever.

I've since learned that constipation is a blockage of the bowels. The colon's job is to absorb water from the waste, bundle it up, and move it along by muscular contractions on its way toward the rectum. The obstruction can be caused by a number of things, including lack of exercise, dehydration from drinking too much alcohol, lack of fiber, and pregnancy. I was guilty on all counts, except the last. Several months of a rice and fish diet probably contributed heavily to my condition.

Unfortunately, the doctor was not fluent in English. Carefully explaining the nature of my problem, I gesticulated vigorously to make sure he understood. The doctor smiled and nodded knowingly as he reached for a container of tiny black pills in the medicine cabinet. "Ah so," he commented as he handed it to me.

The directions on the back of the pillbox were in Japanese characters, but there were two small western-style numbers I could read: 64 and 24.[1] I eyeballed the package carefully as I rotated it looking for any other clues. I asked the doctor if this was the correct dosage as I interpreted it, "I should take sixty-four pills every twenty-four hours?" He looked startled, but responded affirmatively, "Hai, so desne." That seemed like a lot of pills to me, too. After three months on ships I could speak some, so I repeated that question in Japanese, "So, watashi wa taberu roku ju yon korera no mainichi?" He shifted in his chair, clenched his teeth and said with conviction, "Hai," a simple yes.

This horse-size dosage supported my previously mentioned doubts. I don't know what I was expecting, but my ears perked up and nostrils flared. Our conversation continued. "OK, then . . ." I probed. He answered, "OK," while pushing back his chair, effectively ending the discussion. His face twitched ever so slightly. Was that the flicker of a stifled guffaw? As I exited his office, I closed the door behind me, and lingered in the hallway for a minute to see if he would burst out laughing. Instead he inexplicably hummed a few bars from a popular Japanese melody, the haunting cherry blossom song, "sa-ku-raa, saa-ku-raaa . . ."

In retrospect, I now believe that there might have been a misunderstanding between us rather than an intentional act of hostility. The instructions could have been meant as a warning *not to exceed* sixty-four pills every twenty-four hours.

When I got back to my stateroom, I peered down my nose at the canister thinking, "Well, I'm not stupid." I carefully counted out just sixteen of the devilish little pills and swallowed them. They were tiny and lustrous black and smelled like coal tar. I remember being anxious for the outcome. I think I waited several hours before I took sixteen more. It seemed like forever. Nothing happened. After a couple more hours, I took another thirty-two pills to complete the dosage. Even then, except for some hopeful grumblings to break the interlude, there was no progress, but I could feel their dark presence inside me.

In a moment of inspiration it occurred to me that the dose must be sixty-four, but all taken at once. In biology, some activators have a gradual cumulative effect and others have a certain threshold that has to be attained before things start to happen. Of course! I just needed to cross the threshold. Like they say, a little knowledge can be dangerous. I remember carefully counting out the pastilles and organizing them into piles of ten. Then I gulped down a full dose.

Without describing the brutality of the result, I'll just say that I crossed that threshold. The impasse was breached with a shocking suddenness and severity.

Some days later when I was able to crawl out from my cabin again, I was pale and feeble. For some reason the crew seemed to be in unusually good spirits, jovial and happy to see me. Maybe it was part of a modern rendition of omiyage. I had accepted gifts and kindness but had not given back my allegiance to the company. I had not done what was expected of me, so I paid my penance by performing a symbolic hara-kiri. The tension on the ship was remedied.

After my replacements arrived, I left the *Keiko Maru* to board a Japan-bound supply ship. Several of the crew and officers stopped by my cabin to apologize for the way they treated me, confiding that they followed orders. They wanted to express that they were my friends but had obligations to the company. Looking back, I see that this aggregative behavior was a survival mechanism. There is safety in numbers. Group cohesiveness from herds to flocks is common among many animals. They form a superorganism with a different purpose than the individuals within it.

John Steinbeck, also a Salinas native, had a similar tale of industrial fishing. While he and Ed "Doc" Ricketts were exploring the Sea of Cortez on the *Western Flyer*, they encountered a Japanese ship dredging the sea, literally wasting everything in the water column while dumping back all but the shrimp. He said of the Japanese fishermen, "We liked the people on this boat very much. They were good men, but they were caught in a large destructive machine, good men doing a bad thing." It seems that in all human societies there is a struggle between individual expression and the collective action of a group, whether that is a company, an army, or a rioting mob. *What is good for the group is good for us.* Steinbeck described this as capitulation to the group. The concept of safety in numbers may violate our personal integrity, but obedience of the parts is a survival mechanism of the superorganism.

In our case, the mistake we made as observers was to try to correct the institutional corruption from within, while we were a part of the ship's community. While this tactic may work in loosely organized groups, it is less effective when there is a strong hierarchical structure, as in a company. Action is separated from results by many layers of protection. The company had too many ways to exact revenge upon us and to make our lives miserable. What we should have done was to make detailed observations, back them up with documentation, and report those to a higher level in the US government. As I discovered, the individual is often powerless to fight the system from within the structure of the organization. I was like a small bug that can easily be crushed by a larger organism, and only a few people would notice my absence.

Although my relations with the crew had been cured, the fishing company hadn't yet finished punishing me. When the supply ship tied up at the docks in Hiroshima, a company representative met me there and drove me to a posh hotel. He told me to order an expensive dinner with a bottle of wine and to enjoy a massage, all at the expense of the company. I did all that and more to celebrate my freedom from the confines of the ship's iron hull.

The next day I returned to Tokyo for debriefing. The fisheries attaché of the US Embassy waited for me. I was surprised to find him white-fisted and quivering with anger. The company had wired the bill for the evening in Hiroshima to the embassy. The enraged bureaucrat complained, "and how am I going to explain this massage!"

The fishing company had the last laugh. At least they could chuckle to themselves until my report to the agency was submitted. The next year their quotas for crab in the Bering Sea were slashed by 30%. I had learned my lesson.

I wasn't keen to go to sea again, but by the next winter I ran out of money. Once again I found myself in the Bering Sea, this time aboard a Japanese pollock factory trawler, the *Soyo Maru*, during one of the coldest winters on record. And so my relationship with pollock began. The *Soyo* towed a large midwater trawl through the water fishing just off bottom, and the net would arrive back on deck with many thousand pounds of pollock, where it was dumped into a holding bin and sent through a series of conveyer belts into the factory deck below for processing. Checking the accuracy of catch reporting in this type of operation was difficult, and we were working out the details of estimating catches and monitoring bycatch in the trawl fishery. This time I observed and filed my report when I got home.

The year after my second trip, Congress passed legislation to squeeze the foreign fisheries from our coastal waters, and the American rush for white gold followed close behind. Over the ensuing years, the Alaska pollock fishery in the United States alone has grown to an annual gross value of over $1 billion to the privileged few allowed to harvest it. Tremendous battles have been fought over pollock and fortunes have been made and lost.

Even after my time in the Bering Sea, it still hadn't sunk in that I would become a fisheries biologist. I recall my thoughts a few years earlier in college as I sat listening to a well-known, bearded, and bespectacled professor of ecology (Joe Connell) lecture on how beetles bury elephant dung in the African savanna. He was so full of enthusiasm it was almost contagious. We could even calculate a nifty budget of energy flow for them! I remember gazing out the window toward the beach and musing: *The weather is fine, the surf is up, and what am I doing inside listening to this drivel about elephant dung? It has nothing to do with my future.*

One thing led to another, and I started my graduate studies in marine fisheries biology. Biologists try to figure out how living systems work, from cells to ecosystems. Now I realize it was a natural path for me. As a youth I would exasperate my father by taking electronic gadgets apart, leaving them in a pile on his workbench. Biologists are naturally curious about what is inside and how things work, not necessarily how to put them back together or construct them. That minor detail differentiates us from engineers.

Ironically, there was to be more dung in my path. One of my first jobs as a biologist was to analyze fish otoliths (ear stones) from sacks of sea lion scats and spewings collected from their haul-out sites on land to identify the species and sizes of the fishes being eaten. "Spewings" is a word that lights up the eyes of field biologists. It's great because you can get information on diet before it has passed through the digestive system rather than afterwards (that is, the scats). The topic of "What do you do?" at a cocktail party started off some spirited conversations, usually held at arm's length. Most people don't really understand biologists. It was at about this time, I realized that I had metamorphosed into one.

A couple of years ago as I looked back at the detailed journals I kept on the *Keiko Maru*, I realized the transformative nature of my experience on the ship. The questions I had read in textbooks became real. They occupied me for the rest of my career. These aren't novel questions; every ecologist asks them. But I had come to them myself through direct experience. How are the animals of the sea distributed and what controls their abundance? Are the great variations we observe due to harvesting too many or due to changes in climate? What is the role of population density in regulating numbers? Are the stories of overharvested fisheries repeating themselves?

# 1 Introduction: White Gold Fever

What makes a good fisherman? Boldness enables him to take risks and intuition guides him in making quick decisions. Thick calluses and strong hands help with the hard work on deck. Generally, these qualities of a good fisherman make for a good businessman as well, and fishing for pollock was to become good business for a young Norwegian named Kjell Inge Røkke.

One wintry day in 1980, a trawl was deployed from the *Arctic Trawler* into the dark waters beneath it. The fishing master Erik Breivik, fresh from Norway, was looking for cod near the bottom. The net spun off the reel until it was in the water, then the doors, large metal door-shaped slabs that hold the net open from the sides like sails, entered the water with a splash. The winch hydraulics moaned in a loud and low "brrrrmmm" as the thick wire line was played out and the net descended toward bottom. An hour later the trawl was reeled back in. Now the hydraulics screamed under the load. The crew eagerly awaited the riches coming up; millions of pounds of fish were there for whoever had a boat and the nerve to venture into the Bering Sea to catch them.

A great catch, fifty thousand pounds of Alaskan groundfish scooped from the bottom of the ocean, neared the surface. Icy sea spray stung unshaven faces, and the smell of fresh fish hung in the air. Hungry gulls circled overhead. A deckhand shouted out, his hand shot up in a closed fist—the universal sign to "halt"—and the *Arctic Trawler* shuddered from stem to stern, her winches stopped dead. The catch of fish bulging out from the net was so enormous, it split open like a zipper and fish spilled out. The net trailed behind for a hundred feet, undulating in the ocean swell. If the huge bag of fish was hauled up the stern ramp, fish would gush out from the wound in the net—sending most of the catch back to the sea. A body flashed across the deck and leaped from the stern railing onto the rising and falling codend. The crew watched with dropped jaws as the gutsy fisherman clutched the webbing and scrambled in a crabwalk to the open gash of the net. As he bobbed up and down, and stitched the hole, he knew that a slip into the cold sea was almost certain death.

This time the young fisherman Kjell Inge Røkke was lucky, the net was mended, and the catch was saved. The winches roared back to life and one imagines him riding the bucking trawl back up the stern ramp onto the deck in triumph, waving his woolen hat, the crew cheering—a modern day cowboy. A Norwegian cowboy. John Sjong, the skipper of the *Arctic Trawler*, said of the incident, "The next thing I knew he had a couple of needles in his pocket and he ran back on the bag and he sewed it up and saved the tow—he was a hundred feet behind the boat sitting on a bag of codfish." He thought, "Jeez, I wouldn't have done that."[1]

Two years later, Røkke, a high school dropout from Norway's western coast, bought his own trawler and fished for pollock in the Bering Sea. He teetered on the edge of bankruptcy several times and then bounced back to found American Seafoods, eventually controlling 40% of the pollock harvest. Røkke came to play a small but important role in the story of the Alaska pollock fishery. He built on his daring and success as a fisherman, gathered riches in the business world, and made it to the pages of *Forbes* magazine's list of the world's wealthiest people. In two decades, he went from a deckhand on a fishing boat to a net worth of $3 billion. Røkke had come to the New World and realized the American Dream. His companions on the *Arctic Trawler*, Sjong and Breivik, had similar success fishing for pollock. Together they personify the opportunity that the pollock resource presented: great risks were taken and great wealth was gained. These names and others will weave through the narrative of this story.

Alaska pollock is the shy little sister of the cod fish. Not too many people recognize her, but over the past thirty years she's grown up. Now Alaska pollock makes the cod look like a poor cousin. As a seafood product, these days it is ubiquitous. Pollock has been an important foodstuff in Japan since the 1960s; the minced meat is pressed and gelatinized into a product called *surimi*, and the valuable pollock eggs are used in sushi. Surimi can be further processed, adding texture, coloring, and flavoring, to resemble crab and shrimp meat, or made into a sausage called *kameboko*.

When the Alaskan crab populations crashed in the early 1980s, many Americans and Europeans were unwittingly introduced to pollock disguised as imitation crab. But now the market for pollock has matured. The white flesh not only is sent to market as fake crab, but also is fed as a healthy fish product to US soldiers in Afghanistan and to American children in school lunch programs. Pollock is everywhere. It is the pure white meat in fish sticks bought at Walmart and Filet-O-Fish burgers ordered in McDonald's.

The current demand for Alaska pollock outstrips the supply. Pollock is the most lucrative marine fish harvest in American waters and comprises about 40% of the US catch. Since 1950, a total of 187 million tons of pollock have been mined from the waters of the North Pacific Ocean (see Figure 1). The fishery in the United States alone has an annual gross value of over $1 billion. Alaska pollock has alternated with the Peruvian anchoveta as the world's largest fishery since the 1980s, and because the anchovy is primarily reduced to animal feed, pollock continues to be the largest fishery for human consumption.

But has the white gold rush ended? Compared with the 1990s, the fishery certainly has changed, maybe even matured. This modern day buffalo has been domesticated, the pie has been sliced, and the modern day cowboy has been gentrified.

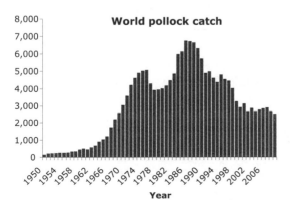

Figure 1.1. The world catch of Alaska pollock (in thousands of tons) showing the dramatic rise in the early 1970s. With the collapse of the anchoveta stock off Peru in about 1972, pollock became the world's largest fishery. It remains the world's largest human food fishery.

## What Do We Know about Pollock and Why Is It So Important?

The scientific name of Alaska pollock, *Theragra chalcogramma*, derives from the ancient Greek words "Ther," or beast, and "agra," or food. Besides being commercially important, pollock is the dominant species in many subarctic coastal ecosystems across the North Pacific Ocean, and hence plays an important ecological role. Befitting its name, pollock is a key prey for beasts of the ocean, marine mammals such as seals and sea lions. They are also prey for seabirds and large piscivorous fishes, such as halibut and cod. Sometimes there are periodic massive die-offs of seabirds that are blamed on shortages of juvenile pollock. The endangered status of some marine mammal populations has been attributed to either a lack of pollock or their poor nutritional value as mammal prey.[2] Juvenile pollock are even an important prey of adults. During certain seasons as much as 80% of the diet of adult pollock in the eastern Bering Sea is made of smaller juvenile pollock.

Pollock is a major predator of lower trophic (food chain) levels and a competitor of small fish and jellyfish for plankton. Cyclic fluctuations in the abundance of pollock have been linked with climate "regime shifts" that are sometimes associated with changes in the structure of marine ecosystems and with dramatic increases in jellyfish abundance in the Bering Sea.[3]

On the surface Alaska pollock may seem like dull and ordinary fish, but under the water they are masters of their environment, capable of living in a variety of habitats from nearshore eelgrass beds to the open oceanic waters of the Aleutian Basin. They thrive across a vast geographical extent that ranges from Puget Sound to the Chukchi Sea, and across the Pacific from Siberia to the Sea of Japan. Pollock survive in a temperature range of one to twelve degrees centigrade, which is fairly broad for northern fishes.

Compared with some modern fishes with lineages originating as early as 250 million years ago, pollock is a recent arrival on Earth. It evolved and has lived in the North Pacific Ocean for around three million years. But that is old when you think of it—humans have been on the North American continent for only about fifteen thousand years. The pollock evolved from an ancestral cod that migrated across the Arctic Ocean during an interglacial period. After some millions of years of evolution in the Pacific Ocean, another warming period allowed some individuals to make a reverse migration back into the Atlantic, giving rise to the Alaska pollock's other close relatives *Theragra finmarchicus* and *Gadus ogak*. The Alaska pollock is genetically so similar to Atlantic cod that some ichthyologists have suggested that it be renamed *Gadus chalcogramma*. Surprisingly, it is more similar to Atlantic cod than to Pacific cod.

Fish populations don't freely wander the world's oceans. Most, like pollock, are structured and organized into geographically based stocks. Some stocks may exchange individuals through migrations and straying, and others can be isolated and over time become genetically distinct. In pollock there are a number of known genetically distinct stocks, and others where there is some limited exchange of individuals among them. As we develop more sophisticated tools to examine the genetic differences, higher resolution of stock structure is obtained. But among scientists studying and managing the populations, there are the lumpers and the splitters—that is, those who believe in vast roaming and mingling schools of fish, and those others who conceive of a multitude of locally adapted populations. The need to manage stocks on the scale of their own natural structure is a key aspect to fisheries management that is still mostly unappreciated.

Why are pollock so valuable to harvest? The short answer: volume! But there are other characteristics that make it easy to fish and catch, and qualities that make it a good product. First, the pollock occurs in large dense schools off bottom that are almost exclusively pollock. This makes it easy to catch large quantities with relatively little effort. Since they are caught off bottom, there is relatively little dragging of the net on bottom, and little disturbance to the bottom habitat. And because the bycatch of other species is low, sorting and processing the catch is easy.

Pollock can be made into a number of food products. The eggs are highly valued in Asia. The objective of the winter fishery is to catch fish while they are aggregated in large spawning groups, but before the eggs swell with water and are released. The flesh of pollock is white with a low fat content and a relatively low parasite load, making it good for fillets and for minced meat to use in fish sticks and fish burgers or to process into surimi. The liver has a high oil content that

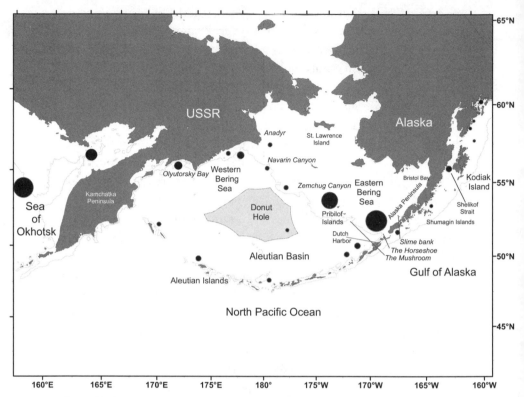

Figure 1.2. Known spawning concentrations of Alaska pollock in the eastern Bering Sea and adjacent seas (round dots roughly proportional to size). There are also spawning areas around northern Japan and extending down the coast of North America to Puget Sound. The lettering indicates some popular names of fishing areas.

is extracted to make omega oils and the remaining waste is made into meal, largely used as chicken feed. The only part of the fish not used presently is its skin.

### Are Pollock and Other Marine Fishes Overharvested?

We expect fish stocks to naturally fluctuate with changes in climate. Usually, during periods of warming or cooling, fish stocks residing at the polar ends of the population's range increase or decrease accordingly, while the reverse happens at the other end. But in the case of pollock, several stocks at either end of their range in the North Pacific Ocean have tended to decrease recently, leading to suspicion of overfishing as a contributor to the demise of several stocks. The roles of climate change versus overfishing in stock declines is an ongoing debate.

The situation is different in the eastern Bering Sea, where pollock is considered one of the world's best-managed populations owing to unusually stable levels of commercial harvests. It is sometimes referred to as the poster child of marine fisheries management. Is it the management of this stock that's different, or maybe is there some attribute of the fish there that makes them more resilient to the pressure harvesting puts on them? In most cases, fisheries managers know that in spite of their best efforts, fisheries ebb and flow as their target populations cycle through periods of high and low abundance.

It isn't as though all is calm in the pollock fishery of the stormy Bering Sea. In 1992 Ross Anderson of the *Seattle Times* wrote of the Bering Sea pollock, "The industry is careening toward collapse."[4] But it didn't happen. In fact, the pollock stock increased, peaking at an all-time high a mere three years later. In 2008 another steep decline seemed imminent and the warning flags were again hoisted up the yardarm. The director of the Greenpeace Ocean Campaign, John Hocevar, said, "We are on the cusp of one of the largest fishery collapses in history."[5] Hocevar added, "Unsustainable fishing rates have been allowed to continue" and "while the fishing industry and others continue to cite the pollock fishery as a model of fisheries management, the pollock population has declined sharply in recent years." But federal stock assessment specialist Jim Ianelli responded, "I don't think we are overfishing." Once again the stock didn't collapse. Then in 2009, news articles in the *Economist and Science* magazines expressed more alarm on an international scale about the health of the major stock of pollock that lives on the eastern Bering Sea shelf.[6] Quotas were decreased to their lowest levels in 30 years by fisheries managers. Has the death knell for pollock sounded?

It seems as if we can't decide, or at least can't agree, whether pollock stocks are collapsing or thriving. Why is it so hard to tell how many fish are in the ocean? And why is it so difficult to manage them? Will the pollock have the last laugh as we try to regulate catches and to engineer the population's natural cycles?

At this moment, marine harvest fisheries, including that for pollock, are on the tipping point of revolutionary change. Fisheries managers are under intense scrutiny. Highly publicized articles, often written by researchers sponsored by the environmental and anti–industrial fishing Pew Foundation, warn about drastic fisheries declines and predict that nothing will remain in the world's oceans but jellyfish in the near future. On the other side of the aisle, scientists, sometimes supported by the fishing industry and government, argue that there is very little overfishing and stocks are recovering. The National Marine Fisheries Service (NMFS) points to the number of US stocks that are recovering from overfishing. The Marine Fish Conservation Network claims that they are juggling the books by pooling overfished and healthy stocks.[7]

A study led by a Pew Foundation Fellow named Dr. Boris Worm concluded that most of the world's fisheries would collapse by 2048.[8] Other prominent marine scientists have disagreed with this assessment of the state of the world's fisheries. One of the world's most highly respected fisheries scientists, Dr. Ray Hilborn at the University of Washington (who has been described as a "hired gun for the fishing industry"[9]), said of Worm's projection, "It's just mind-boggling stupid."[10] Worm replied that it was a "news hook to get people's attention."[11] Then in a later paper, Worm, now partnered with Hilborn himself, seemed to reconcile their differences (once again supported by the Pew Foundation).[12] More recent studies indicate that fish stocks aren't as bad off as we once thought, but about 25% of them are still overexploited or collapsed.[13]

Other reports have indicated that there is a broader ecosystem consequence to overfishing and that we are "fishing down" the food chain.[14] This term means that as the more delectable top predators, such as tuna, swordfish, and salmon, are fished out, industrial fisheries then start fishing for the smaller animals that are the prey of the larger fishes. Thus, human fishers become both highly efficient predators and competitors as well. Recently, that conclusion has also been attacked.[15]

In response to charges of overfishing, the At-Sea Processors Association said, "Could the 'gloom and doom' crowd have it wrong? University of Washington professor Ray Hilborn thinks so." The report continues, "Dr. Hilborn's work . . . showed that fish stocks, in fact, are increasing in abundance."[16] The word bombs that are lobbed back and forth show the extent to which we are engaged in a major battle for the public's attention and favor. Scientific findings announced in press releases reflect the powerful forces struggling behind the scenes. When there is an economic stake, whether that is related to corporate profits or environmental fund-raising, it has the appearance of agenda-driven science to me, analyses done for the purpose of finding a specific result.

Large industrial ships may, or may not, be fishing down the food chain as a consequence of overfishing, but some are fishing lower on the food chain purposefully. Small pelagic crustaceans, such as krill, and small fishes, such as anchovy, concentrate valuable marine oils. The omega-3 and omega-6 oils are important in a healthy diet and are in high demand. Norwegian companies are harvesting krill in the Antarctic using new "green technology."[17] In a public relations bid to change their image, they call their trawlers "Life Science Factories." By the way, guess who presides over the company that is competing with whales and penguins for krill? It is Kjell Inge Røkke from the *Arctic Trawler*, who now rules over a maritime empire from Norway.

## Who Owns the Fish in the Sea?

The human demand for seafood has simply exceeded the capacity of the oceans to produce it. With that in mind, it is frightening to consider China's emerging demand for seafood and the ensuing pressure that ocean resources will face in the future. Solutions that are sometimes proposed are to privatize the ocean's resources or to transform the Earth's seas from hunting grounds to marine farms.

Some years ago, an analyst with Prudential Securities said, "Fish are the last of hunted animals, it's the last of the great commodity ventures."[18] Fishermen themselves have been described as the last true hunters of the world. Fisheries managers wish their charges would behave more like farmers, who plan ahead and preserve part of their crop to ensure the seeds needed for next year's harvest. However, the sad fact of history has been "Fishermen reap but they do not sow."[19] When a fisherman talks about good business and the money to be made in harvesting the sea, there is an eager yet wistful look his eye that hints of a primal ancestor, a hunter closing in on his prey and anticipating the feast. In this case, the feast is the potential riches to be harvested, and they can be enormous. One pollock fisherman described his best day fishing in a small trawler. He caught 500 tons.[20] At a value of about $300 per ton that's worth $150,000. Not bad for a day's work. But he remarked it was worth a lot more than that because it was prime roe. In 1992 the *Chelsea K*, fishing in the Bering Sea with a net spreading out 38–42 fathoms and 8–14 fathoms high, caught 590 tons of pollock in one tow. Fishing is like a treasure hunt. You put your gear in the water to see what you will catch. Sometimes the gamble pays off big. In 2008, a herring fisherman near Sitka, Alaska, caught nearly $250,000 worth of herring in a single thirty-minute set.[21]

Our concept of limits to the ocean's resources is undergoing a revolution. Management of natural fisheries is under increasing scrutiny. A few short years ago, harvests from the ocean were considered boundless. Now even the industry points out that they are not. As recently as the 1970s, the harvests from the sea were still considered as having nearly unlimited potential. Important and respected fisheries scientists proclaimed that fish not harvested over and above the level needed to reproduce themselves were wasted in the ocean. The *sustainability* of a fish population was defined as the ability to maintain itself under harvesting pressure, or even to maintain economic viability. Now we are very much aware that other animals in the sea compete with humans for fish as prey, and the word *sustainable* has an ecosystem context. More and more, we realize that what humans catch is robbed from the mouths of other species.

Who owns the fish in the sea? Most people would be surprised to know that the pollock pie has already been sliced. The fishery has been closed to new fishermen, and ownership of this resource has been transferred from the public to *catch quotas* (there are other terms that are basically synonymous, including *individual transferable quotas, individual fisheries quotas, catch shares, rations,* and *cooperative quotas*). Among economists and politicians, privatizing the resource is one of the popular solutions to overfishing, while also a means to increasing profits. It's also been popular among fisheries managers, administrators, and some lucky fishermen who are the recipients of the public largess. Are the fish in the sea "common property"? Does the public "own them" or do the fishermen? Can the government grant private access, can they sell rights to the fish, can they even exclude access? These are hard questions that have deep social and legal connotations.

There are rights or privileges attached to the quota holder to catch a certain percentage of the harvest. In the case of pollock, these are "attached to the steel," or in other words, each ship is assigned a quota (sometimes by legislation, sometimes by a management agency mandate, and other times by a cooperative). The owner of the ship can choose to "lease" his quota and make millions of dollars without ever dipping his net into the sea. In fact, some vessels have not fished for pollock for years while their owners reap the rewards of their early investment in a public resource, one that most people think belongs to and is managed for the benefit of all Americans. Not only has the pie been sliced, but also the pieces have been passed around the table.

Wally Pereyra, a savvy businessman, chairman of the pollock fishing company *Arctic Storm*, and a prominent figure in the industry, said, "The days of being able to go from one resource to the next are gone, there aren't any uncharted horizons left."[22] Ironically, Pereyra's son is building an aquaculture farm in Guatemala to grow tilapia.[23] Right now the biggest competitor of Alaska pollock for the whitefish market is farm-raised tilapia and catfish from Asia. Some farm-raised fish has even made it to the market sold as Alaska pollock. In England, something called "Young's Flipper Dippers" was labeled as containing Alaska pollock, but actually contained Vietnamese river cobbler, a freshwater catfish. An astounding 264 million portions of fish products were reported to be potentially mislabeled in one year.[24]

As the world shifts from hunting the sea to farming or ranching it, one is reminded of the cultural transformation in the West during the change from hunting buffalo to ranching and farming. The story of the West is one of conflict between wild herds and introduced ones, farmers and ranchers, ranchers and herders. Freedom versus fences. We can envision similar conflicts and clashes between sea harvesters, sea ranchers, and sea farmers. Already farm-raised salmon exceeds the supply of wild salmon, and controversy rages about the effects of farm-raised and genetically engineered fish on the remnant wild populations. These issues include competition for prey, transmission of diseases and parasites, and loss of natural genetic biodiversity.

## Pollock in the Context of History

The boom and bust nature of the Alaska pollock fishery parallels similar episodes in the much older California sardine and Atlantic cod fisheries. Will pollock share the same fate of overexploitation and devastating collapse? Will stock declines set off enormous economic and social consequences to local fishing communities and stress the marine ecosystem? Pollock seem to be rocking back and forth between increasing and decreasing trends, similar to what happened to the sardine population in the 1930s and 1940s. Then in 1946 the sardine population collapsed drastically but academic scientists predicted a recovery. There was a minor and brief resurgence, before the California sardine population finally died with a whimper. It took another thirty years for the California sardine to recover, and even then, not to its prior levels.

The cod stocks off the coast of Canada also bounced up and down, with ensuing controversy about their status and the strength of incoming year classes. Then in the late 1980s, they collapsed. Just before the collapse, one very bright and responsible scientist, who was at the time the chief of research at Canada's Department of Fisheries and Oceans, "cheerfully" went on television saying that "lots of young cod were coming through, several year classes of them." These young cod never made it.[25] This scene plays back eerily similar to what we now hear about pollock.

In his landmark book *The End of the Line,* Charles Clover said of the northern cod collapse, "An army of scientists in one of the world's wealthiest and most advanced nations managed to destroy one of the richest fisheries in the world, while convincing themselves for a decade that they were doing no such thing. The Newfoundland cod collapse was the nightmare that shook the world out of its complacent assumption that the sea's resources were renewable and being managed in an enlightened manner."[26] For the world's fisheries scientists, and the Canadians are among the smartest and most innovative, that is a painful criticism. But there is another side to this story. I met the former chief of research mentioned above at a cocktail party not long after the collapse. He'd been transferred to the west coast and away from the headlights in Newfoundland. He talked openly and at length about his frustrations. I was spellbound by his monologue of how politicians and bureaucrats ignored the advice of biologists and then pointed fingers elsewhere for the demise of northern cod.

There are also important differences between the cod and pollock fisheries. For one thing, the Atlantic cod fishery is hundreds, if not thousands, of years old. That is a lot of history under which layers of traditional fishing culture have been established. Fishermen from

Galicia took cod from the Grand Banks of eastern Canada even before the time of Columbus, and the Vikings were there before them. Politics and social pressures have made it difficult to stop such deeply ingrained tradition. In the northern cod collapse, Spanish fishermen exceeded the recommended catch quotas for cod on the Grand Banks by seven-fold.[27] Now Spanish fishermen are subsidized by the European Union not to fish in their own waters and they dominate the high seas fleet of fishing pirates who pillage the open ocean, harvesting illegal, unreported, and unregulated fish.[28]

By contrast, the commercial pollock fishery is only thirty or forty years old in the northeastern Pacific Ocean. While cod and sardine fishermen reminisce about what happened years ago, pollock fishermen are riding the roller coaster of this natural population's ebbs and flows right now. The pioneers in the fishery, such as Kjell Inge Røkke, Erik Breivik, Chuck Bundrant, John Sjong, and Wally Pereyra, weave in and out of the pollock story, and some of them are still actively involved in the fishery. Management of the eastern Bering Sea pollock stock is promoted as an example of good fisheries management, "the Cadillac of fisheries management." Pollock in the eastern Bering Sea and Gulf of Alaska have been certified by the Marine Stewardship Council (MSC) as being sustainably harvested.[29] But Cadillacs are expensive, and the infrastructure needed to manage the fishery is pricey.

Some of our problems managing fisheries can be directly linked to political decisions made just after World War II and to the treatment of marine fisheries as a commodity to trade on the market of international treaties. In the 1970s, fish quotas for the Japanese fishery in US waters were used as leverage to open Japanese markets to US exports.[30] This history of fisheries as a commodity, rather than a living resource, might partly explain why the National Marine Fisheries Service (NMFS) is part of the Department of Commerce, a situation which causes many to wonder how it fits in there.

Historically, the role of fisheries management has been to develop and sustain the economic viability of the fishery.[31] Because of NMFS's association with commerce and fisheries development, many conservationists see federal fisheries scientists as tools of the fishing industry.[32] On the other hand, NMFS has the final say in pollock harvests, and the agency maintains the proud record of never allowing the harvest quota for this species to exceed the recommended levels set by science. This is not the case for many other fisheries. Why is the history of pollock different?

What follows is the story of the development of the pollock fishery in the backdrop of science, politics, policy, and economics. How did the fishery develop? Who fished for pollock? How was the resource managed? An important aspect of understanding the pollock fishery is recognizing what aspects of the natural history of Alaska pollock make them so very desirable to fish while at the same time make them resilient and yet vulnerable to overfishing. In examining and criticizing how the fishery has been managed, it is important to recognize what we knew about the ocean and its fish resources at the time. Were management decisions made within the context of the best available science?

The pollock fishery and its management evolved within a bigger backdrop of human understanding of our interaction with the ocean. What have been the conflicts over huge amounts of fish removed from the sea, both in terms of environmental stress and economic development? The rapid development of the fishery led to new paradoxes and improvisations for new management structures that surfaced abruptly. Unforeseen ecological problems and social clashes emerged. Great wealth was retrieved from the ocean. But is the wealth of the sea unlimited? What follows is the untold story of the Alaska pollock fishery, a rush for white gold in the cold dark sea it inhabits.

# 2 A Historical Background: From an Inexhaustible Ocean to the Three-Mile Limit

It's ironic that the founding father of the "unlimited potential and freedom of the seas" died as the result of a shipwreck. In 1609, twenty-six-year-old Hugo Grotius published *Mare Liberum* (The Free Sea),[1] a small book with a great impact on the relationship between the ocean and society. Grotius, also known in the Netherlands as Huig de Groot, was a child prodigy from an educated family. He entered the university at age eleven, received his PhD when he was fifteen, and published his first book in Latin by the time he was sixteen.

Is it far-fetched to link the writings of Hugo Grotius in the early seventeenth century to the modern pollock fishery? No, because how we think about the sea today, how we determine fisheries harvest levels and who can take them, is based on this history. The United Nations Convention of the Law of the Sea evokes the principles established by Grotius when it states that the high seas are open to all states, with freedoms including navigation and fishing.

The story of Grotius is interesting. When he wrote his book, Grotius represented the Dutch East India Company as a jurist. The Dutch Republic had granted a twenty-one-year monopoly for trade in Asia to the company. The Dutch East India Company defined commercial innovation; it was the world's first megacorporation and the first to issue stock. The company possessed quasi-governmental powers to wage war, negotiate treaties, and establish colonies. It was paving an inroad into the Portuguese Empire's sphere of influence in the spice trade.

At the same time, Europe was experiencing turbulence in politics, religion, and economics. Spain and Holland fought as opponents in the Eighty-Year War in which Portugal and Spain were allied. Spain reached the pinnacle of its power during this "Age of Expansion." Besides fighting the Dutch, the Spaniards were battling the forces of the Ottoman Empire to the south, the English to the west, and the Aztecs and Incas in the New World. Holland itself roiled in chaos, with internal battles between Catholics and Protestants, and in-fighting among Protestant factions.

The Dutch East India Company engaged Grotius to defend its seizure of a Portuguese merchant ship and cargo. At the time, the seafaring nations were busy exploring the oceans and establishing lucrative trade routes. In particular they established a burgeoning trade with Asia. The Eighty-Year War with Spain had begun as a struggle about religion, but ended up involving commerce and trade domination. A truce had just been negotiated in the war, but Portugal and its ruling state Spain monopolized the East India trade, and Dutch merchants wanted a piece of the action.

In his defense of the Dutch East India Company, Grotius laid out the argument that the sea is free to use by all for travel and trade. His ideas were based on the concept of natural law, whose content is set by nature and is thus universal, as opposed to laws of custom and convention. However, even in the times of Grotius, deep lines were drawn between separate schools of thought: is the nature of man based on his own self-interest, or does he serve the common good of society?

Grotius reasoned that because the sea is free flowing, it is more like the air than the land. No one owns it since it cannot be enclosed. Because the sea and the resources it contains are limitless, it cannot be a possession of anyone but is common property for the use of all—whether it is for navigation or fisheries. In the particular case of the Dutch East India Company, because Portugal was denying free access to the sea, then by natural law it was fair to capture their ships and cargos as just punishment.

The premise that the sea and its resources are inexhaustible, and that the sea is fluid and ever changing, formed the basis of the new concept of "the freedom of the seas." However, the English and Scots considered Grotius's book an assault on their claim of exclusive fishing rights in the North Sea. The Spanish viewed it as an attack on their overseas empire.[2] In spite of these powerful voices, eventually the freedom of navigating the oceans and harvesting the animals within took hold and became "the law of the sea." The law evolved over time with regard to coastal zones. Coastal countries extended their claims seaward from land to three miles offshore on the basis of the distance a cannon shot could travel to protect sovereign property.[3]

Grotius is recognized as one of the founders of the modern school of natural law and even as an important contributor to American political thought. In spite of his lasting achievements, fate treated him roughly in his own lifetime. Grotius actively participated in philosophical and theological discourse, and found himself on the losing side of an argument between the ruling Calvinists and a reformist movement about what today seems like a minor issue involving religious freedom. The ruling Calvinists arrested Grotius and sentenced him to a life in prison. He was lucky. One of his unfortunate colleagues was beheaded. With the assistance of his wife, Grotius escaped from his dank cell by hiding in a chest of books to be smuggled out. He lived in exile in Paris, then moved to Sweden and later became their ambassador to France. On a trip home to Sweden he was shipwrecked. He washed up on shore in the Netherlands, where he sickened and later died of his weakened condition.

Since the time of Grotius, the extent of the sea and the abundance of animals it contains have become synonymous with the word "boundless." Most nations established the tradition of freedom to fish the seas up to the three-mile limit. In the later part of the twentieth century, this view changed drastically. Nations sought more enclosure of their adjacent seas, and the existing perceptions of ocean productivity and how (or even whether it is necessary) to manage marine resources shifted. We can better understand the present-day conflicts of resource development by knowing something about the history of thinking about ocean productivity and fisheries management. Standing back to look at the historical landscape puts the developing pollock fishery in a global context.

## The Debate on Fisheries

At the Great International Fisheries Exhibition of 1883 in London, two prominent British scientists clashed about the nature of ocean resources. Thomas Huxley and Edwin "Ray" Lankester were both celebrated Fellows of the Royal Society and colleagues. Their argument foretold a stormy future in fisheries science.

Huxley had little formal schooling and was largely self-taught. He had left school at age ten due to family circumstances (his father had lost his job as a math teacher). The young man apprenticed from the age of thirteen to medical doctors. Later Huxley served as an apprentice and then assistant surgeon on the HMS *Rattlesnake*. It was on the *Rattlesnake* that he became a naturalist and comparative anatomist, and formed many of his opinions about the ocean. Huxley later taught as a professor at various institutions, and served as Inspector of Fisheries, and President of the Royal Society. He proved that he had a prodigious intellect and became known as one of the greatest autodidacts of his time. A bombastic character, he often lectured as "Darwin's bulldog." Huxley gained fame for debating religious authorities and coined the word *agnostic*. After all, he was the godfather of Darwin's children and grandfather to author Aldous Huxley.

On the other side of the argument stood Ray Lankester. Lankester was a protégé and admirer of Huxley, but he held different views of the world. His father was a medical doctor, and friends with Huxley and Darwin. Lankester was educated at Cambridge and Oxford. He was a Fellow of the Royal Society, was trained as an evolutionary biologist, and held a position as professor at Oxford. He has been described as England's most influential biologist.[4] Lankester was an ample man with a big presence and warm personality, but he could make enemies with his rude demeanor, and he is said to have become cantankerous in old age.

In the inaugural address of the exposition, Huxley argued, "I believe that it may be affirmed with confidence that, in relation to our present modes of fishing, a number of the most important sea fisheries, such as the cod fishery, the herring fishery, and the mackerel fishery, are inexhaustible. . . . and probably all the great sea-fisheries, are inexhaustible; that is to say that nothing we do seriously affects the number of fish. And any attempt to regulate these fisheries seems consequently . . . to be useless." He continued, "And I base this conviction on two grounds, first, that the multitude of these fishes is so inconceivably great that the number we catch is relatively insignificant; and, secondly, that the magnitude of the destructive agencies at work upon them is so prodigious, that the destruction effected by the fisherman cannot sensibly increase the death-rate." Huxley did concede that "there are fisheries and fisheries [that is, fisheries with different characteristics]," and species like salmon and oysters had life histories that allowed them to be overexploited.[5]

Lankester held the floor in the final summary speech of the London exhibition. He rebutted, "It is a mistake to suppose that the whole ocean is practically one vast store-house, and that the place of the fish removed on a particular fishing-ground is immediately taken by some of the grand total of fish, which are so numerous in comparison with man's depredations as to make his operations in this respect insignificant." In other words, the fishes in the sea are not unlimited, and captured fishes are not necessarily replaced by a ready supply of others from offshore. Lankester believed that the removal of the parents by fishing was going to impact the production of young. He brought up the concept of equilibrium between fish populations and their predators[6] and he voiced an omniscient sense of future concerns, "The thousands of apparently superfluous young produced by fishes are not really superfluous, but have a perfectly definite place in the complex interactions of the living beings within their area." He could have written the last statement as a modern environmentalist.

In the short run, at least for the next century, it seems that Thomas Huxley won the day. According to Tim Smith's analysis, the positive outlook for the future of fisheries presented by Huxley was just too tempting.[7] "An inexhaustible supply of fish" was just what the industry wanted to hear. That isn't to say that scientists weren't aware of the dangers and practice of overfishing, but optimism and the concept of the overriding regulatory control of population dynamics by natural processes were more popular with a muscular industry and tended to influence management decisions. It wasn't until 1949 that the influential English scientist Michael Graham told that the wave of history would wash over Huxley. In his book *The Fish Gate*, Graham wrote, "T. H. Huxley's advice on fisheries has, nevertheless, to be judged critically, and the verdict is that, although he was 100% right in 1866, by 1883 he failed to take account of a new development."[8] That development was the modern trawl and the steam trawler to pull it, which revolutionized fisheries.

The Huxley-Lankester debate was not the first, or last, time the argument about the inexhaustible fisheries of sea was aired. In fact, the effect of fishing versus that of the environment is a lightning rod that today still attracts the spark of conflict and galvanizes fishermen, scientists, and politicians to opposite poles.

The advent of steam trawlers in the 1880s allowed fishermen to drag larger nets and bring in more fish than ever before (see Sidebar 1). In 1900 the English biologist Walter Garstang entered the fray and warned about overfishing the ocean. He pointed out the increase in fishing power due to the steam trawlers. Garstang had a rather blunt approach that was disagreeable to the other British scientists but much appreciated by the Scandinavians, who considered him an honorary Viking.[9] In 1900 he published an article titled "The Impoverishment of the Sea," in which he showed that catch rates of plaice in the North Sea decreased, in spite of increasing fishing effort, presumably due to overfishing by trawling. His work showed that specific popular fisheries, such as plaice, are exhaustible.

Michael Graham was the director of England's Ministry of Agriculture and Fisheries' Lowestoft Fisheries Laboratory from 1945 to 1958. Graham was an eccentric character. He used to ride his horse to and from work in Lowestoft; for the ride home at night, he wore a bowler hat with a red light affixed to the port side and a green light on the starboard.[10] Graham insisted that his scientists at Lowestoft be called "naturalists."[11] In *The Fish Gate*, Graham said, "The trail of fishery science is strewn with opinions of those who, while partly right, were wholly wrong."

Graham showed that North Sea fish stocks increased after a lapse in fishing during World War I, and therefore harvesting had an effect on fish populations. Furthermore, they could recover. The findings of Graham and Garstang seemed to tip the balance of the Huxley-Lankester debate toward the latter.

Graham later directed a group of his scientists at Lowestoft to look at the effect the reduction in fishing during World War II had on fish populations. British landings of marine fishes had increased dramatically until World War II, when they dropped precipitously for the period of the war because of mining and torpedoing of the high seas. In 1941 a committee of scientists met to discuss what measures to take after the war to ensure that the North Sea fish stocks would not be overfished again. They assembled several forward-looking proposals, including regulation of the number of vessel-days spent fishing, tonnage regulations on the fleet size (limited entry), and minimum mesh sizes in the nets to decrease the harvesting of small fish. Unfortunately, the political process after the war diluted these recommendations. The science was overruled and the stocks were soon overfished again.[12] Fishing effort increased, fish populations decreased, and from the 1950s onward catches declined. Looking forward, landings in 2000 were far less than they had been 100 years before. Removals of fisheries surpluses were not insignificant.

Similar arguments were popping up in other seas. On the West Coast of the United States there were the locally well-known Thompson-Burkenroad "debates" on Pacific halibut. William Thompson's work on exploitation of halibut had a strong influence on Graham. Thompson believed that fisheries populations had been overfished, and demonstrated that the relaxation of fishing in the Pacific halibut population resulted in population increases. Martin Burkenroad questioned Thompson's conclusions and suggested that the environment was a major factor influencing fisheries population dynamics. Thompson and Burkenroad never actually debated face to face, but they wrote a series of papers with opposing viewpoints.[13] Today there is no definitive resolution to the Thompson and Burkenroad disagreement. The waters around the issue remain murky.

Although fisheries scientists recognized problems in local fisheries, on a more global scale the riches of the sea were being proclaimed enthusiastically. It was widely believed that fisheries held the key to solving world hunger.[14] If local fisheries became depleted, we just had to look at other species and in other locations.

Rachel Carson, the "Earth Mother" of the environmental movement and author of the landmark *Silent Spring* and the beautifully written *The Sea around Us*, started her career as a young aquatic biologist in the US Fish and Wildlife Service.[15] In 1943 she wrote about the need for conservation of natural resources in her pamphlet on seafood, "Food From the Sea," but her solution was to exploit other underfished species in order to relieve pressure on the most popular ones.[16] In other words, fish more but diversify. She wrote that from the standpoint of human welfare, thousands of pounds of fish go to waste in the sea each year simply because we aren't catching them. A few species are overexploited, while there is wasteful underexploitation of other species. Carson's work in these early years was likely influenced by her patriotic sense of duty in the spirit of the war effort, and the wartime push to find new sources of protein.

## A Science of Fisheries: The Beginning Years

Fisheries research is an applied science and has always been driven by economics and politics. The "golden age" of fisheries research in the late nineteenth and early twentieth centuries was fueled by the increasing industrialization of marine fisheries and a growing awareness of the sea as a resource. Nowadays the pressure is on fisheries research to end overfishing, to manage on the basis of ecosystem considerations, and to forecast the effects of a changing climate in order to sustain the economic viability of the industry.

The mysterious nature of the ocean attracted a public curious to explore it in late nineteenth-century Great Britain. In the spirit of the times, the Frenchman Jules Verne published his book *Twenty Thousand Leagues under the Sea* about the adventures of the exiled Captain Nemo aboard the *Nautilus*. The public's imagination was captured.

The prosperity of the Victorian era funded ocean exploration to satisfy public and scientific curiosity. At the time, the Azoic Theory was *en vogue*, which considered that the deepest regions of the sea were devoid of life. However, early exploratory fisheries surveys had trawled up some bizarre life forms, which were publicly displayed. The HMS *Challenger* expedition in 1872–76 was put together to explore the deep oceans.[17]

The expedition also had some very practical commercial underpinnings. In 1850, the Anglo-French Telegraph Company laid the first telegraph cable across the English Channel. This was followed by laying of the first successful transatlantic cable in 1866. Submarine cabling was dominated by English companies and provided an economic boom, facilitating transmittal of news, shipping information, military orders, and government business with a speed that was inconceivable a few short years before. Knowledge of the bottom topography, locations of trenches and obstacles, would be of enormous help in planning the placement of thousands of kilometers of expensive cables. The depth soundings involved in the exploration of new pathways for transocean cables provided a great economic benefit.

The *Challenger* sailed with 243 scientists, officers, and crew, and journeyed 127,670 km around the world. The news of the expedition stoked even more interest in marine resources around the world.[18] Within a short time new marine research programs and laboratories started up in Italy, Russia, Scotland, Canada, England, Germany, Denmark, Japan, the Netherlands, and the United Sates.

The Norwegian government took a strong interest in ocean science because of the importance of fishing to the well-being of its people. In 1864, the government tasked G. O. Sars to examine the factors causing catches of cod around the Lofoten Islands to fluctuate wildly, which had drastic effects on the local economy. According to Smith, this was the beginning of an international push to study the scientific production of fisheries.[19]

During the period from 1860 to 1920, scientists, and particularly those in northern Europe, made great advances in marine science providing the basis for understanding population dynamics and laying the groundwork for scientific management of harvests. A young Norwegian named Johan Hjort and other Scandinavians, including C. G. Petersen, F. Nansen, G. Ekman, M. Knudsen, B. Helland-Hansen, and O. Pettersson, began studies of oceanography and fisheries interactions in order to explain changes in the distribution and abundance of marine animals. Almost every tool and concept we use in understanding fish populations today was developed during this period.

Meanwhile, in Germany, F. Heincke demonstrated that there were local stocks of herring in 1875. Methods for quantitatively sampling plankton were developed by V. Hensen in the 1880s. His research group in Kiel worked out the foundation of the plankton production cycle in the ocean. The Danish scientist C. G. Petersen developed methods for tagging fish to study their migrations. He also introduced the concept of density dependence and the use of fish scales to determine ages. He applied this latter technique to demonstrate that there are yearly cohorts of fish.

In 1894, Hjort inherited Sars' leadership of Norwegian fisheries science. He thrived in this role and grew into an icon of marine science. Hjort was known for his energy and his volcanic temper. Pictures show a stout, square-headed man, with wire-framed glasses and a chevron mustache that was fashionable at the time. Hjort started his career in medical studies but he switched fields to study zoology, which led him to the University of Munich where he got his doctorate degree in 1892.

One professor described Hjort this way: "As a superior, he was without peer; helpful, kind, patient—as an equal, rather difficult because he always thought he was right—and as a subordinate, sure of himself and full of the desire to oppose."[20]

Hjort was uniquely capable of integrating the most recent findings from the diverse fields of oceanography, population dynamics, and fisheries science. In the beginning of the twentieth century he built upon the foundations created by earlier European scientists to come up with concepts of why fisheries populations fluctuate. His ideas are still widely respected and cited today.

At the time, the livelihoods of many early twentieth century Norwegians depended on the abundance of fishes, mainly cod and herring. Drastic changes in abundance were thought to be caused by large-scale migrations of fish schools in and out of the fishing grounds. Hjort instead proposed that the abundance of fish populations changed with the successful recruitment of new year classes. That is, highly variable survival of newly born cohorts into local populations was driving the great fluctuations. He also developed ideas about how the survival of the earliest stages of marine fishes changed with environmental conditions, and how the survival rates contributed to year-to-year changes in availability to the fisheries.

Hjort was an ardent supporter of commercial fisheries and set a precedent for fisheries scientists to invent ways to improve the methods of finding, catching, and processing fish.[21] His efforts to modernize fisheries and to increase catches in order to improve the Norwegian economy mixed well with his idea that natural variations in ocean conditions are the main cause of fluctuating fisheries. Hjort, like Huxley, believed that there was no need to limit fishing, and if anything, fishing should intensify. Ocean conditions, not fishing, caused depletion of fisheries. In fact, Hjort was an outspoken adversary of anyone who thought that overfishing was a problem.[22] At the time, Hjort was at least partly correct because the technological level of fishing off the coast of Norway was relatively primitive. The fishermen had an enormous amount of ocean whose resources were at their disposal to harvest. If fish were in short supply locally, the fishermen just moved to another place.

Hjort exported his theories of natural variations of marine fish populations. Eager to share his ideas of improving fisheries with the rest of the Western world, he proselytized through a series of courses that he taught in Bergen, as well as by trips to Canada and the United States. His efforts to evangelize led directly to the belief, still held by many fisheries scientists, that research on fisheries development—methods to catch more fish and to do so more efficiently—was more important than work on conserving fish stocks.[23]

With the coming of the Great Depression in the late 1920s and through the duration of World War II, politicians lost interest in funding marine research, and resources were shifted away from the field because of the worsening economic situation.[24] Instead, the emphasis in fisheries science shifted toward compiling fisheries statistics and their analysis, as these efforts didn't require expensive expeditions. Increasingly, mathematics was applied to understanding fisheries dynamics.

Outside of the field of fisheries in the discipline of ecology, scientists had shown the importance of such concepts as carrying capacity, competition for food and space, and predation in population dynamics.[25] Theoretical fisheries scientists applied some of these concepts toward understanding fisheries populations. During this "middle age" of fisheries that lasted until about 1970, various methods of describing fish population growth, dynamics, and harvest levels were developed, including the surplus production model and the relationship between spawning population size and recruitment, most notably by William Ricker, Ray Beverton, and Sidney Holt.[26] Although fisheries scientists grasped the concept of density dependence for developing harvesting strategies, there was a raging debate about density-dependent versus density-independent control of other animal populations in ecology.[27] The aspect of density-independent control of fisheries populations has yet to be implemented much in fisheries management models.

## A Global View of Productivity

The pioneering marine scientists of Scandinavia and Europe developed the concept that conditions in the ocean were important to the production capacity of fish populations on a local scale. Later, in the twentieth century, oceanographers had the idea of examining the amount of primary production in the oceans to estimate potential fisheries production on a global scale. Alternatively, some fisheries scientists thought the way to do this was to evaluate current world harvest levels and add that to estimates from unexploited fish populations. Various scientists used one or the other method to report the potential harvests from the world's oceans.[28] The resulting estimates were vastly different. However, the US government chose to integrate the more optimistic view into policy.

Studies of production in the world's oceans were useful to establish the upper limits to fish resources, but on a practical and operational level the estimates were extremely high. One of the major sources of variability is a value used for the transfer efficiency of biomass between trophic levels. The use of 20% rather than 10% for the efficiency translated into a doubling of estimated harvest potential. The problem is that fish don't neatly fall into specific trophic levels, and they change levels as they grow.

In 1964, the US Interior Department reported that "under sound conservation, management of the world's ocean could provide 500 million tons of fish and seafood produced annually."[29] In 1967, the highly regarded fisheries scientist Wilbert McLeod Chapman upped that estimate to 1 to 2 billion tons. Chapman didn't consider that surplus fish production was needed to sustain the ecosystem. He argued that surplus fish that are not harvested, are lost forever.

Also in 1967, Congress authorized the Commission on Marine Science, Engineering, and Resources to study ocean policy. Two years later, the blue-ribbon committee of American scientists filed its report, titled "Our Nation and the Sea," with sections ranging from preservation of shorelines to effective use of ocean resources.

The report addressed marine fisheries production, and estimated that annual harvests of 400 to 500 million tons could be taken from the sea. These estimates presented a very optimistic view of food potential from the ocean. The committee stressed the potential for increasing the number of American jobs in the fishing industry and providing protein to an increasing world population. Chairman Julius Stratton was the former president of the Massachusetts Institute of Technology and director of the Ford Foundation. He was a good friend of Chapman. James Crutchfield, an economist on the commission, said that he himself wrote most of the report, and that the extremely high estimates of yield were not in the report that he authored. He suspected that Chapman used his influence to insert the optimistic wording in the final report.[30]

In 1969, John Ryther of Woods Hole Oceanographic Institution estimated that on the basis of primary production calculations, world fish production was 240 million tons per year. However, he recognized that man must share the resource with other top level predators, such as birds, whales, and seals.[31] Ryther thus calculated that the potential sustained yield of fisheries was not greater than 100 million tons. Now, as conceived by Ryther and even by Lankester almost a century before, we at least acknowledge the prey requirements of upper trophic levels and the desirability of maintaining healthy and sustainable ecosystems.

Compared with estimates from the fisheries sector, Ryther's calculations of sustained fisheries production made over forty years ago are much closer to the mark, we now realize. Daniel Pauly estimates that world fisheries catches have leveled at about 90 million tons per year, and some scientists think this level is too high (the level of marine fish yield, not including shellfish and mollusks, is less than 80 million tons).[32]

By the 1970s, great cycles of changing ocean conditions, as well as shorter term fluctuations, and their impacts on fisheries were beginning to be recognized. The Russell Cycle in the English Channel was identified as a sixty-year cycle of plankton and fish abundance that was related to warming and cooling trends.[33] In the early 1970s, dramatic changes in ocean conditions occurred off the coast of South America, there was a strong El Niño, and the devastating collapse of the world's largest fishery, the harvest of Peruvian anchoveta, became famous.

## Development of Fisheries Management in the United States

In 1871 the US Congress established the US Fish Commission to address industry conflicts. There had been increasing competition and strife among long-lining and trap fishermen for declining local resources. A young professor of natural history at Dickinson College named Spencer Baird proposed studying the issue before enacting legislation, and he was named the first commissioner of fisheries. Baird was unsalaried and worked out of his home. Public antagonism toward the idea of regulating harvests motivated Baird to avoid this conflict. Instead, he developed programs to augment fisheries. In the 1880s, billions of young cod larvae were released annually into coastal waters from American hatcheries. But there was a nagging controversy about the effectiveness of the hatcheries in actually augmenting fisheries. This dispute motivated studies of larval fish survival in the wild. Baird conceived a program to comprehensively study the population dynamics, ecology, and oceanography of marine fishes.[34]

In Norway, Hjort also questioned the efficiency of marine hatchery programs, and debunked Norwegian marine enhancement programs. Although marine hatchery programs in the United States were never shown to be effective, they persisted for almost seventy years until 1952. The concept of marine hatcheries periodically reemerges in US fisheries policy as a solution to overfishing.

After World War II, there was growing confidence in the power of science and engineering. The basic mathematical approaches for managing fisheries had been laid out. In 1947, Ira Gabrielson, director of the Fish and Wildlife Service, declared that fishery biology was an "exact science."[35] Over the next fifty years, mathematical management models to describe the dynamics of populations and to estimate sustainable harvest levels became increasingly sophisticated. Fisheries management is still based on an engineering approach to regulating the resource, which is not too different from harvesting crops and trees.

One of the principal designers of US fishing policy was Wilbert Chapman. He was the director of the School of Fisheries at the University of Washington, and in 1948 became the Undersecretary of State for Fisheries in the Department of State. Later he worked closely with the tuna fishing industry as an advisor. Chapman pushed the concept of maximum sustainable yield (MSY) as a policy rather than a scientific approach. He said, "Less fishing is wasteful, for the surplus of fish dies from natural causes without benefit to mankind."[36] According to Alan Longhurst in his book *Mismanagement of Marine Fisheries*, Chapman knew of the work by the American Milner (Benny) Schaefer on formulating MSY and pushed the concept to acceptance before it was published.[37] Chapman was trying to balance the views of Alaskan salmon fishermen who wanted to limit foreign fishing on their stocks, and tuna fishermen who were themselves fishing in foreign waters. Chapman's solution was that salmon had already been fished to "maximum sustainable yields," but tuna had yet to reach that level and needed to be more fully exploited.[38]

## The Precedent of Politics and American Fisheries

From their very beginnings, US fisheries agencies were put in place to assist the economic viability of the fisheries industry. In 1791 Thomas Jefferson called for helping American fishermen with tax relief, increased tariffs on foreign products, and promotion of markets abroad. The first US Fisheries Commission assisted the fishing industry with subsidies, loans, development of seafood products and technology, marketing, and trade promotion. In 1947 the United States joined the rush to industrialize world fisheries when Ira Gabrielson proposed a postwar program to support the growth of the American fishing industry.[39]

Important legislation impacting fisheries was passed in the twentieth century. Under the Merchant Marine Act of 1920 (Section 27 was known as the Jones Act), vessels landing fish in the United States had to be constructed there, and had to be owned and crewed by Americans. Later, the Fishing Vessel Construction Differential Subsidy Program provided funds to make up the difference to build ships in the United States versus cheaper ships abroad.[40] The Capital Construction Fund, established under the Merchant Marine Act of 1936, allowed tax on profits from vessel operations to be deferred if they were used to renovate, construct, or purchase ships. This fund essentially was an interest-free loan.

In the Fish and Wildlife Act of 1956, a low-interest Fisheries Loan Fund was created and it doled out $31.3 million from 1957 to 1973. Then in 1972, Congress created the Fishing Vessel Obligation Guarantee Program (FOG), which encouraged private lending by guaranteeing loans to vessels in case of defaults.

In 1929 Clarence Birdseye introduced a fast-freezing process to the food industry. It was mainly used to preserve vegetables. Then in 1953 the Birdseye Division of General Foods announced the production of frozen fish sticks. The US government saw fish sticks as an opportunity to improve conditions in American fisheries, and engaged in a program to develop and promote fish products such as these. The companies Mrs. Paul's and Gorton's joined the effort to process and market fish sticks.[41] In 1954 the Saltonstall-Kennedy Act collected custom duties on imported fish products and directed them to programs for marketing and research to help the US fishing industry. Marketing efforts increased with the passage of the Fish and Wildlife Act of 1956.

Fish protein concentrate (FPC) was invented in 1962 and marketed as a solution to world hunger. However, the Department of Agriculture and the Department of the Interior's Bureau of Commercial Fisheries (BCF) and their respective constituents engaged in a war over the competition between agricultural protein, such as vegetable, soy, and milk protein, supported by Agriculture versus fish protein supported by BCF. Rules were even put down on FPC by the US Food and Drug Administration, such as limiting the size of bags in which the protein could be marketed to one pound. This severely impacted the industrial-scale usage of the product. Later, restrictions were put on fluoride concentrations in FPC, which necessitated deboning the fish before processing, driving the cost up. Eventually FPC production halted in 1973.

## The Great Theory of Fishing

Concepts of fisheries management in Europe and the Americas have undergone radical changes in the past 100 years. In response to declining stocks, early fisheries managers thought to replenish populations by growing and releasing massive numbers of eggs and larvae. Up to 2.5 billion cod and haddock eggs were released annually into the Atlantic Ocean by US scientists during the early twentieth century, as well as hundreds of millions released by Norwegian scientists, all without measurable effect. Hjort himself argued against the value of stock enhancement efforts for cod. The tradition of replenishment of depleted stocks has continued with salmon hatcheries, and the effort to enhance other stocks continues to the present day. In some cases, marine enhancement has been relatively successful—for example, that of Japanese flounder, but the effect of such enhancement on natural populations is unknown.

Managing fisheries by regulating harvests is a relatively new concept. When it became obvious that marine enhancement was not working, another approach was needed. In the 1930s, English scientists' studies of plaice in the North Sea and the examination of halibut in the Pacific Ocean by Americans showed that these fish populations were being overharvested. As coastal areas became depleted of these fish, the fishing effort would expand into unexploited areas until there weren't any remaining. The concept of harvesting just enough to ensure that the population could still sustain itself took shape.

The theories of fisheries science commonly used today probably arose, at least partly, from German concepts of scientific forestry developed in the nineteenth and twentieth centuries. This discipline focused on the concepts of optimum and sustained yields and shaped the ideas of early fisheries practitioners, such as Hjort and the Canadian scientist A. G. Huntsman.[42] The influence of forestry can be seen directly in the writing of the influential Huntsman, who drew analogies between trees and fish and their management to maximize productivity. Hjort himself got his PhD in Germany and was exposed to Norwegian forestry practices. He imported his philosophies to Canada through a sustained visit in 1914–15, with Huntsman as his assistant.

There was also a belief that fisheries could be planted and harvested, much like crops, which led to the marine enhancement efforts.[43] A. G. Huntsman, as president of the American Fisheries Society, wrote about managing fisheries stocks as if they were crops. This conceptual view of fisheries persists today. For example, the Korean government has established several areas that they term "marine ranches." These are areas where the habitat is groomed and improved, and seeded with young fish that will be harvested later.[44]

Fisheries scientists were at least aware of, if not involved in, developments in the discipline of animal ecology. Hjort spent the years 1916–21 at Cambridge, where he was exposed to the well-known ecologists Charles Elton and David Lack, and the mathematical models of populations developed by Raymond Pearl.[45] Hjort later worked on harvests of marine mammal populations and developed models to adjust catch rates in order to be in equilibrium with the maximum population growth rate.

Michael Graham also believed that there must be an optimal point somewhere in the relationship between a sustainable population size and fishing intensity.[46] Graham had been studying the impact of trawlers, at that time called North Sea "scrapers." Graham wrote in *The Fish Gate*, "It is evident that free fishing, and free building of trawlers, outran the capacity of the market for that type of fish. If the fish could be better preserved the market would doubtless expand; and that is the direction in which co-operation should go in the future."

Graham made several important contributions to the theory of fishing. He advocated the "logistic" or s-shaped growth curve of populations (where at low densities the fish population grows faster and then at higher densities the growth rate begins to slow until it reaches a maximum at the carrying capacity). The logical conclusion of such a model is that there is an optimal harvest level where the population growth is maximized.

The formulation of the theory of MSY (the highest catch that can be continuously taken from a population without adversely affecting production), based on the foundation provided by Hjort, Graham, and others, was finally put together by Benny Schaefer. In his model, the maximum yield is obtained when the density is thinned so that production is optimized to the highest population growth rate, or about one-half of the virgin unfished biomass (B). Later versions of MSY approximated that MSY $= \frac{1}{2} \times B \times M$. Of course, the estimates of M (the mortality rate due to natural causes, as opposed to that due to fishing, F) are approximate and vary, and the prefished biomass is not known precisely and likely to be highly variable as well. The end result was a crude estimate of MSY. Nowadays, the approach is more sophisticated using models, but still based on the assumption of surplus production. The idea is to optimize the harvest, while leaving behind the largest number of reproducing animals to breed and survive at the maximal production rate. Traditionally, US fisheries scientists have used MSY as a harvest target, while British scientists consider it an upper limit. In most fisheries the biomass at MSY, or $B_{msy}$ is at about 30% of the original unexploited biomass, but many times the level is set at 40%. For Alaska pollock in the eastern Bering Sea, $B_{msy}$ is typically about 31% of the unfished spawning biomass. The $B_{msy}$ level tends to change with each assessment as the parameters change; in a practical sense, it is used as one of several reference points to gauge the health of the stock.

Graham noted a problem and described the "Great Law of Fishing," which is, "As the fishing power increases the stock falls, but the yield at first rises. Later, as we shall see, it ceases to rise, and that creates the main problem of fishing."[47] The fisherman then has "the urge to expand," or increase his hunting territory. Graham realized that "fisheries that are unlimited become unprofitable" and conversely, that imposing limits would restore profit to a fishery.

## The Foundation of NOAA

President Richard Nixon acted on a recommendation by the Stratton Commission's report "Our Nation and the Sea," using Executive Order 1154 to form the National Oceanic and Atmospheric Administration (NOAA).[48] The new agency consolidated marine and atmospheric research in the Department of Commerce. Most of the functions of the Department of the Interior's Bureau of Commercial Fisheries were transferred to the new National Marine Fisheries Service (NMFS) with NOAA. The new agency had jurisdiction over most living marine resources of the nation's oceans. Jurisdiction over marine mammals was split between Interior's US Fish and Wildlife Service (USFWS) and NMFS. The NMFS would manage the more oceanic species, such as whales, seals, and sea lions, while USFWS would manage those with a coastal lifestyle, such as polar bears, otters, and manatees.

At the time NMFS did not have a regulatory role except for mammals. State jurisdiction generally extended seaward to three miles offshore, and beyond that was the high seas. This changed, first in 1966 when Congress extended the US jurisdiction of fisheries to twelve miles offshore, and second in 1976—dramatically—with passage of the Fishery Conservation and Management Act (FCMA). With this legislation the Fishery Conservation Zone (FCZ) of the United States was extended to 200 miles offshore.

This was the status of our knowledge of the production of the ocean and the role of fisheries science as the pollock fishery launched and developed from the late 1950s to the early 1970s. The US government and, indeed, many other governments were enthusiastically pushing ocean fisheries as a way to feed an increasing population on the planet. Although there were naysayers to the policy of unbounded fisheries, and at least the potential to overfish popular coastal species was well known, the prevailing view was that the ocean's resources were practically unlimited. The oceans were underexploited and had the capacity to feed the world. The sea was boundless. "As yet we do not know the ocean well enough. Much must still be learned. Nevertheless, we are already beginning to understand that what it has to offer extends beyond the limits of our imagination—that someday men will learn that in its bounty the sea is inexhaustible."[49]

The role of government agencies was to develop fisheries and to maintain their economic sustainability. Complementing these views, mathematical models were developed to manage the fisheries and were based on the concept of fishing down stocks to a point where their productivity would compensate and then their growth would increase. This increased production could be considered surplus. Fisheries science had reduced the complexity of harvesting the ocean's living resources to a few equations. Scientists and administrators brimmed with confidence. Around the world, governments encouraged and assisted fisheries to grow.

·  ·  ·  ·  ·

SIDEBAR 1: EARLY TRAWLING IN THE BRITISH ISLES

*Men have been dragging net bags across the bottom of coastal waters since Egyptian times. One of the first modern-type trawls was the "beam" trawl, so named for the beam of oak or ash mounted on the upper lip of the bag to keep it open, while rocks were attached to the lower lip to submerge it. The first official reference to the beam trawl was about 1370 when Edward III was petitioned to ban it from the Thames Estuary for the damage it had done. It was then known as the "wondyrchoun" or wondrous device. It was ten feet long and one can only imagine how difficult it was to hoist aboard the heavy beam by hand. The beam trawl persisted in some fishing communities around England until the 1930s.*

*In the late nineteenth century dragging converted from sail power to steam power with the screw propeller. Wood hulls gave way to iron and then steel. In fifteen years sailing trawlers were a thing of the past in the major fishing ports of Hull and Grimsby. Coinciding with the arrival of the steam trawler, the otter trawl was invented. This first mention of this type of gear was 1865. The mouth of the net was spread open by two iron-clad wooden doors (known as otter boards), each with its own wire.[50] As the ship moved forward, the doors spread out and upward, opening up the net. The bottom was held down by a weighted line and the top held open by floats. It needed* a lot of power to pull and worked best with steam boats. By 1890 the beam trawl gave way to the otter trawl. Pockets of sailing smacks survived in the small fishing ports that served local populations. Lowestoft had a sailing fleet until 1939. The boom of the steam trawl era from 1900 to 1914 also gave rise to a new cuisine—fish and chips. Lower-quality product kept at sea longer could be used in fish and chips because it was masked with batter. Just before World War II there were 25,000 fish-and-chip shops in England.

The arrival of steam power in England also led to dual-purpose ships. Converting to steam involved a greater capital investment, meaning that ships needed to work year-round to pay for it. Drifter-trawlers fished for herring with drift nets and trawled for bottom fish. David Butcher noted, "Since the introduction of the steam engine in the 1880s there has been a dangerous kind of in-built snakes and ladders game in the trawling industry. It works like this: bigger engines and better trawls catch more and more fish for a time, but after a while catches fall because of the depletion of stocks. Then, in order to maintain fishing at previous levels, the size of the engines and the weight and spread of the gear has to increase—which again will eventually cause a further drop in catches. And so it goes on."[51]

In the 1920s came the pioneering technique of filleting fish. Around this time ships started converting from coal to diesel. Coal was increasing in cost owing to strikes in 1921 and 1926 that led to shortages. Plus, diesel fuel was more compact, so ships could carry more and roam to more distant waters.

•　•　•　•　•

# 3 Fishing the High Seas: Japan and the Soviet Union Develop the Alaska Pollock Fishery

*The fear and dread of you will fall on all the beasts of the earth, and on all the birds in the sky, on every creature that moves along the ground, and on all the fish in the sea.*                                                     Genesis 9:1–3

On September 2, 1945, the USS *Missouri* was anchored in Tokyo Bay. Onboard the warship, General Douglas MacArthur and representatives of the Allied Powers and Japanese government signed the document ending World War II with the surrender of Japan. After the great devastation of the war, General MacArthur and the Supreme Commander of Allied Powers (SCAP) had arrived in Japan to find a massive food crisis. Japan's economy had deteriorated in the last two years of war. Cities were bombed out, harvests were poor, and the food distribution network was disrupted. The infrastructure of the country had been broken and their great fishing fleet was all but destroyed. Starvation was knocking at the door. Where would the Japanese people turn to relieve their misery? To the east was the vast Pacific Ocean. To the north and west lay the socialist threat of the Soviet Union. Just beyond the horizon loomed the rising tide of the Chinese Communist Party, which was winning its war against the Kuomintang (Chinese Nationalist Party).

Fishing the sea has always been important to Japan, and the practice is embedded in the traditions, culture, and art of the Japanese people. As far back as the Jomon period, 10,000 BC to 300 BC, the Japanese were a fishing and hunter/gatherer society.[1] To solve her modern problems, Japan returned to the sea. Partly as a result of that decision, their postwar recovery turned into an economic miracle. Within a decade they became the world's greatest fishing nation. In another decade they developed the little-known and undervalued Alaska pollock into the world's greatest fishery. The story of the early development of the pollock industry is largely that of the postwar fishing industry of Japan.

Today, seafood accounts for about 40% of the protein consumed by the Japanese people. Japan is currently the second largest fish-consuming nation in the world after China, consuming 7.5 billion pounds of fish each year, or about 10% of the world's harvest. The average Japanese person eats about seventy kilograms of seafood a year; their nearest rivals, the Scandinavians, consume about half as much.

This enormous consumer demand makes Japan the largest importer of seafood in the world.[2] Because of Japanese dominance of the global seafood market, trends in seafood consumption in Japan have a great effect on the international marketplace. Seasonal supply and Japanese demand for seafood fluctuate, such as for the New Year's holiday when fishcakes are a popular item. Buyers find supplies tighten as Japanese producers of fish cakes purchase supplies in preparation for the holiday. The state of the Japanese economy also plays an important role in seafood pricing, such as in 2009 when the economic crisis of Japan impacted the purchasing power of Japanese for expensive seafood. The demand for surimi fell; wholesalers had plenty of product in their freezers, and prices dropped by 20%.[3]

The commerce of seafood has a deep history in Japan. The nature of fisheries there changed from subsistence to a trading economy during the tenth century. This commerce developed rapidly in the Edo period (1601–1868) with the advent of more modern fishing gear, the descendants of which are still in use today.[4] Fishermen were able to capture fish well beyond their own needs. In the Edo period, Japan began exporting seafood to China. To obtain more fish, the harvesters needed to go beyond the coastal waters of Japan. Because of Japan's dominance of global fisheries, the rise of high seas fishing in the Pacific and of global industrial fishing is largely the story of fisheries development in Japan.

## Rise of Japan's Distant Water Fisheries

Japan's first major expansion beyond their own coastal waters was in the 1700s off Sakhalin for salmon. By 1868, extensive fisheries had developed in the Russian Far East.[5] In the Russo-Japanese Treaty of 1875, Russia exchanged the Kurile Islands for Sakhalin and some fishing rights, and Japan obtained permits to establish shore plants to process salmon, which they maintained through the Russian Revolution and up until World War II.

The large fishing companies that characterize Japan's industries have roots going back to the 1880s. One of the largest companies, Taiyo Fishery Ltd., was founded by Ikujiro Nakabe in 1880 as Hayashi Kane Shoten.[6] The other Japanese megacompany, Nippon Suisan Kaisha Ltd., was founded in 1911 as a fishery department within the Tamura Kisan Company.[7] These companies provided the capital and infrastructure to support the far-ranging high seas fisheries that developed.

The Japanese government encouraged looking to the sea as a source of protein in a resource-limited land. After a law promoting distant water fisheries was enacted in 1897, distant water exploratory surveys started around Japan and Korea.[8] This was followed by the Pelagic Fish Encouragement Act of 1905 to assist building large motorized ships.[9] In the early 1900s Japanese fishing boats were plying the waters of the Sea of Okhotsk and the western Bering Sea.[10] Between 1895 and 1941 Japan was expanding its resource harvests across the eastern Pacific from Sakhalin to the Philippines and into the Indian Ocean.[11] By the 1930s Japan dominated the fisheries of Southeast Asia.

As early as 1930, Japan had floating canneries in Bristol Bay processing crab.[12] As we shall later see, the presence of crab fishing vessels there in the early 1930s was a key point in the forthcoming development of Japanese fisheries in the Bering Sea. There was also a limited trawl fishery beginning in 1933. However, the fishing company Nippon Suisan reports that they sent the first trawler into the Bering Sea even earlier, in 1929.[13]

In 1936, Japan began a study of Pacific salmon resources in Bristol Bay, which caused an uproar in US fisheries politics because salmon were already intensively harvested by American fishermen. They were afraid of Japanese interception of salmon on the high seas before American fishermen had the opportunity to catch them when they migrated inshore.

In 1937, the US government sent a note to Japan indicating that there was evidence of salmon interception without the authority of the Japanese government, raising a serious problem and creating considerable friction between the two countries. Then in 1938, a provisional settlement was reached. Japan halted scientific investigation of the salmon resources and declared it would continue the policy of not granting commercial licenses to fish salmon in the region; however, "without prejudice to the question of rights under international law," meaning that the issue wasn't permanently settled.[14] The presence of Japanese fishermen in the eastern Bering Sea prompted the US Congress to find out something about the resources in the sea off the Alaskan coast, and to fund exploratory fisheries for king crab and bottomfish, but these were delayed until after the war.

By 1939, Japan was the world's largest fishing nation, with some 10,000 vessels over ten tons in size.[15] Fisheries were also a major source of jobs, employing some 1.5 million citizens in industry-related work. World War II interrupted the fishing industry, because many boats were converted for wartime use, its workforce was depleted, and the freedom of the seas was restricted by enemy ships. By the end of the war, Japan's fishing fleet had been destroyed by bombs and torpedoes, and their shoreside plants were severely damaged. The fishing capacity of the fleet was reduced to 40% of the prewar levels, and fishing rights in international waters were restricted.

## Post–World War II Development

The Potsdam Declaration of 1945 issued on July 26, 1945, set the conditions for ending the war with Japan. Those terms were harsh: "Following are our terms. We will not deviate from them. There are no alternatives. We shall brook no delay." The declaration, issued by the United States, Nationalist China, and Great Britain continued, "The alternative for Japan is prompt and utter destruction." The declaration did imply a precedent for future fishing rights. "Japan shall be permitted to maintain such industries as will sustain her economy.... To this end, access to, as distinguished from control of, raw materials shall be permitted." As history tragically unfolded, Japan rejected the terms, which led to unleashing the massive destructive powers of the atomic bombs dropped on Hiroshima on August 6 and Nagasaki on August 9 of 1945. However, the terms of the Potsdam Declaration set a precedent, establishing future rights of Japan to reestablish a high seas fishery.[16]

With Japan's surrender after the nuclear bombing, the occupying force of the Supreme Commander of Allied Powers (SCAP) and his staff arrived in Japan to find a shortage of food and the threat of widespread starvation. The SCAP was in the awkward position of both meeting the immediate needs of the people and finding a way to quickly make the economy self-sustaining.[17] General MacArthur said, "SCAP is not concerned with how to keep Japan down, but how to get her on her feet again."[18] The solution was found in the fishing industry. Although the shore-side industries and fleet had been damaged, the potential ship-building capacity of postwar Japan was enormous.

The SCAP made the strategic decision to restore the shipyards with the purpose of developing Japan's fishing capacity. Under the authority of General MacArthur, Japanese entrepreneurs began to rebuild the coastal fleet.[19] The SCAP authorized the conversion and repair of vessels and increased production of new vessels in shipyards. It gave the fishing industry priority status in obtaining fuel and other resources.[20]

The SCAP had other reasons for developing Japan's fisheries and improving the living conditions in Japan. They were concerned about the growing influence of the USSR and communist China in the region. The US government was nervous about socialist activity in Japan, fueled by long lines for rationed food. Japan was seen as a potential capitalist outpost that could establish a foothold in Asia.

In an expanding fishery, Japan's fishermen were at first restricted to fishing in their own territorial waters. But the harvesting capacity soon outgrew the resource and fishermen strayed outside the boundary. Koreans were complaining about illegal Japanese fishing activity in their territorial waters.

The area that Japanese boats were allowed to fish was restricted by the "MacArthur Line," which initially delimited the coastal waters surrounding Japan. As the capacity of the fishing fleet continued to grow and resources were depleted, pressure to expand the area of fishing increased. After all, Japan had the freedom of the seas before the war, and had a lot of distant water experience. Gradually, the fishing zone was allowed to incrementally expand over the seven-year period of Allied occupation. The SCAP's policy favored industrialization and exploitation rather than conservation.

The war had left a major legacy for fisheries in the greatly enhanced technology that was created by wartime research. The development of LORAN (Long Range Navigation), radar, and sonar assisted in safe navigation, position finding, and locating fish. The development of strong and lightweight synthetic fibers allowed larger nets to be deployed. Older nets were made from cotton twine, which was weaker and more likely to tear, and which soaked up water, making them heavy.

Improved propulsion made ships faster and more powerful. They could travel farther and catch more fish. Refrigeration allowed them to keep their catch from perishing and to develop distant water fishing. All of these technical innovations paved the way for expansion and globalization of marine fisheries. In the two decades after World War II, global marine fish catches increased 300%. The other important development after the war was investment by governments into fishery capacity, led by Japan, the USSR, Norway, and the United States.

The SCAP decided that Japan must play a role in managing their high seas fisheries and the Japan Fisheries Agency was formed. In 1950, the Diet (Japan's legislature) passed the Marine Resource Antidepletion Law. The agency was now vested with the power to regulate activities of the fishing fleet for conservation.[21]

The SCAP's policy of rebuilding the fishing fleet's capacity and extending the high seas boundary met international opposition. Australia and New Zealand didn't want the Japanese ships returning to their coasts. Alaska and West Coast fishermen feared overfishing by the Japanese and increased competition.[22]

Significant US opposition to SCAP's initiatives to expand Japanese fishing came from the Department of the Interior, which was concerned about Alaskan interests, and also from Wilbur Chapman, Special Assistant to the Undersecretary of State for Fisheries and Wildlife. MacArthur criticized Chapman as a mouthpiece for special interests, but Chapman continued to oppose policies that would allow Japan free and equal access to the high seas as provided in the Potsdam Declaration.

With the approach of the signing of the Final Treaty of Peace with Japan (also known as the San Francisco Peace Treaty of 1952; signed in 1951 and ratified in 1952), the granting of sovereignty status to Japan and the exercise of their historic freedom to fish the high seas were viewed by some as a threat to the fisheries of other nations. The Law of the Sea at that time was the three-mile limit of national sovereignty and freedom of the seas outside that limit. The US State Department perceived the need to negotiate fisheries issues before signing the treaty, and convinced the SCAP to grant temporary sovereignty to Japan for that purpose.

The International Convention for High Seas Fisheries of the North Pacific Ocean was negotiated in 1951 before the signing of the peace treaty. It was signed in May 1952, just after the peace treaty, and came into force in June 1953. The Final Peace Treaty stated that Japan would negotiate with the Allied Powers for agreements on the conservation and development of fisheries on the high seas (Article 9).

The High Seas Treaty established that under international customs and laws, the three signature countries, the United States, Japan, and Canada, would have the right to exploit the resources of the high seas, with the goal of ensuring the maximum sustained productivity of the fishery resources of the North Pacific region.[23] However, there was an important inclusion in the treaty. A clause called "The principle of abstention" changed the whole concept of the Law of the Sea and the three-mile limit of sovereignty with the freedom of all use for all outside that limit. The clause indicated that the treaty countries would agree on a conservative regime to limit fishing rights on the high seas when certain conditions were met—that is, when there were already fully developed domestic fisheries of the adjacent country. Specifically, Japan would not fish for halibut, salmon, sardines, or herring on the western seaboard of North America.

Chapman had pushed for a 150-mile extended zone of jurisdiction in the treaty, but the California tuna industry wanted a three-mile limit so they could fish the coasts of Central and South America. Chapman jumped ship and supported the three-mile limit. Also, tuna was originally on the list of species whose fisheries would be regulated by an international commission, but by now Chapman was working for the tuna industry, and he was able to get tuna excluded from the treaty.

The San Francisco and High Seas Treaties finally closed the door on World War II, and critically allowed Japan to fish in areas of the North Pacific where they had exploited resources before the war. Since Japan had fished in the Bering Sea as early as 1929, this vast sea was included in the area where Japan could now harvest resources.

Over the next few years, Japanese boats began to fish in areas they had used before the war. However, several nations did not participate in negotiations or did not sign the San Francisco Treaty, including Korea, China, and the Soviet Union. The USSR and Japan were locked in disputes over repatriation of prisoners of war, the jurisdiction of the Kurile Islands, and fishing rights.

The USSR had entered the war against Japan late and did not join the end-of-war treaty negotiations. In 1945, the USSR unilaterally took over all the Japanese fishing facilities on Sakhalin Island, the Kurile Islands, and Kamchatka Peninsula. After the war the USSR, along with Korea and China, seized hundreds of vessels and thousands of fishermen caught in the disputed waters. Finally in 1956, a compromise was reached between Japan and the Soviet Union normalizing relations, and they signed a bilateral fisheries agreement over the North Pacific.

## USSR's Postwar Fisheries

The fisheries of the Soviet Union had been hit even harder than those of Japan by the war. At the end of World War II, the Soviet fisheries industry was almost completely destroyed. Almost 5,000 fishing vessels were wrecked and harbors, ports, and shoreside facilities were destroyed. Afterwards, the USSR, along with Japan, embarked on one of the most dramatic and rapid fishery expansions the world has ever seen.

The USSR was assisted in their fisheries development by a Lend-Lease arrangement with the US government, by which their vessels were fitted out.[24] But it was the massive investment and centralized plan of reconstruction by the Soviet government that made it happen. Soviet officials realized that agriculture alone wasn't providing all the protein needed by its people, and looked to the potential of the seas to fill the gap. The Soviet government began to rebuild the fishing industry, and by 1967 the USSR had the world's largest fleet of fishing and associated vessels. However, a simultaneous massive reconstruction of navy ships meant that their own shipyards were full, and many vessels were produced in Poland and East Germany. Now with its huge fleet, the Soviet fishery needed to expand into the high seas to meet the harvest targets set by central planning.[25]

World War II left numerous legacies for Pacific fisheries. Many geopolitical tensions existed around fishing. Japan and the USSR had disagreements over fishing that went back to the Russo-Japanese War of 1904–1905. The United States wanted to strengthen the Japanese economy as a bulwark in Asia against the communist governments in China and the Soviet Union. The Soviet Union supported a twelve-mile territorial sea, while both Japan and the United States wanted a three-mile territorial limit. For the United States, the three-mile limit was founded in the freedom of the seas, which allowed the movement of military vessels as well as fishing boats on the high seas. Restrictions on where boats could fish were seen as setting a precedent that might lead to other restrictions. But it was alarming when Soviet factory processing ships showed up on the Pacific coast of the United States at the height of the Cold War in 1966, fishing legally just outside the three-mile corridor. Domestic fishermen, coastal communities, and the fishing industry spent the next decade seeking to set a 200-mile limit and to control foreign fishing in coastal waters of the United States.

## The Early Years of the Pollock Fishery

Japan planted the seeds of a large-scale industrial fishery for pollock, initially by their exploratory fishing in the Bering Sea, and then through the industrial production of surimi. At present, Japanese fishermen play a minor role in the global harvest of pollock, but the consumer demand for pollock in Japan is still a major factor in the market place.

Pollock have been used as a food by the Japanese for centuries. The indigenous people of Hokkaido Island used dried pollock.[26] The species was fished around the Japanese archipelago and along the northeastern coast of Korea, and was mostly prized for its roe, "the golden eggs."[27] The Japanese people didn't particularly like the meat, but because of pollock's abundance it was an important commercial fish.[28] Pollock meat loses its "freshness" quickly, and old pollock develops a particularly bad taste. This probably contributed to pollock's unpopularity among the Japanese prior to the development of cold storage systems.

Akira Nishimura, a well-known Japanese expert on pollock, relayed his father's impression of the fish: "After World War II, they served unfresh pollock every day for rations, and after that I still hate pollock."

The first industrial pollock fishery started in 1903 at Iwanai Bay on the Sea of Japan in the Hiyama region. Before that, it was known to be abundant around Hokkaido Island but was mainly a bycatch in the cod fishery. The pollock fishery began because of declining harvests of herring. It was little valued but this changed when a demand for dried pollock emerged in China and Korea. Longline and gillnet fisheries developed around 1910 to 1920; a trawl fishery was formed in the 1930s, and a peak in catches was observed in 1936 at 140,000 tons.[29] Large quantities were harvested for their roe and a lot of the meat was wasted, or used in the production of fish oil and meal. By this time almost all the local fishermen were engaged in the pollock fishery. Then in 1952–53, the pollock around Japan disappeared. It was a devastating blow to the local fishermen and they had to migrate to other areas to fish in winter for the next fifteen years.[30]

In contrast with the Japanese, Korean people historically appreciated pollock and it was commercially valuable there before the development of surimi. There are reports dating back to the seventeenth century of Koreans eating pollock eggs.[31] The annual average prewar consumption of pollock by Korea was 250,000 tons.[32]

With the postwar development of industrial fisheries, the Japanese fishery for pollock expanded into the Bering Sea and also in the waters along the coasts of Kamchatka and the Kurile Islands. Kasahara reported large catches of pollock made by Japanese mother ship trawl fisheries in the eastern Bering Sea before 1960.[33] The growth of the pollock fishery in the Bering Sea accelerated when the yellowfin sole (*Limanda aspera*), which formed the basis of a large fishmeal industry, declined sharply in 1964 and was replaced in the industry by pollock.

Although pollock had been an important food resource to coastal northern Asia for centuries, it was not used off the coast of North American by native fishermen or by industrial fisheries. In the United States, fishermen were interested in higher value products, like halibut and crab, and in profitable nearshore fisheries for salmon and herring. There was a general lack of interest in groundfishes (fishes that live near the bottom of the water) in the eastern Bering Sea because of their low economic value, the travel required to catch them, and the lack of shoreside facilities.[34] Historically, a fishery for cod had been established in Bristol Bay in 1864, and previously US fishermen had been fishing for cod in the Sea of Okhotsk from 1850–1880.[35] But these fisheries did not survive long.

As early as 1888, American fishermen knew that groundfishes were plentiful in Alaskan waters, but they didn't know just how abundant they were. "The fishing grounds [in Alaska] are believed capable of furnishing an unlimited amount of cod."[36] Marketing has always been a major impediment to fisheries development on the West Coast, and there was no market for pollock; thus, what was known about pollock was "the Alaska pollock is one of the best baits known for cod."[37] Little did they realize what was going to develop in the future. From 1955 to 1975, American groundfish fisheries in the North Pacific were limited to small-scale trawling, and halibut and sablefish line fisheries.[38]

The other issues for American fishermen to consider in harvesting groundfishes of the Bering Sea included the harsh weather and the remoteness of the place. There were also the considerable problems of preservation, transportation, and marketing of an unknown product that they would have to confront.

However, by the late 1950s the Japanese were certainly aware of the importance and potential of pollock, leading one scientist to say, "There is no doubt that this is one of the most widely distributed and most abundant fish species in the whole North Pacific."[39]

### The Magic of Surimi and Expansion of the Fishery

The industrialization of a fish paste called surimi stimulated the growth of pollock fisheries across the North Pacific. Surimi was first developed as a food product in Asia over 1,500 years ago, but not until 1960 was the industrial process to make surimi invented.[40] From the beginning of the industrial process, pollock has been a preferred species for making surimi because of its low lipid content, which allows the congealing of proteins to make the gelatinous substance, and because of its bland taste and pure white flesh.

In the making of surimi, the fish is headed and gutted and the meat is skinned and deboned. Then the flesh is minced, rinsed, and pulverized into a jelly-like paste. From the fish paste, surimi is further processed into fishcakes (kameboko), imitation crab, fish sausage, and other products. The advent of the technology for making surimi led to the explosive rise of the industrial fishery for pollock.

At first, the great increases in pollock fishing occurred in waters surrounding Japan, but these stocks were limited. By 1958, Japan was in the Bering Sea and fishing for pollock, but mainly for use in making fish meal. The eastern Bering Sea groundfish fishery had a combined catch of 103 million pounds in 1958 but rose to more than 1.6 billion pounds by 1961 when pollock began to be used for surimi.

In the eastern Bering Sea, the pollock fishery experienced explosive growth. Between 1964 and 1971 catches of pollock increased ninefold.[41] The catch peaked in 1972 at 1.6 million tons. American fishermen didn't pay much attention to this development itself since the Japanese trawl fisheries were a long way from trawling grounds off the coast of Oregon and Washington, and they didn't compete in the same market. As the foreign groundfishing increased in the eastern Bering Sea and Gulf of Alaska, the main concern of American fishermen was not about competition, but about the bycatch and depletion of the fisheries they were interested in, halibut and crab, and the interception of salmon headed back to Alaskan coastal rivers to spawn.

In the 1960s, restrictions designed to limit catches of pollock in the Bering Sea were not an issue. But that changed in 1972–73 when US negotiators insisted on restricting Soviet and Japanese fisheries. Finally in 1973, a bilateral treaty with the United States put a quota on Japan's pollock catch in the eastern Bering Sea of 1.5 million tons. However, the quota was self policed and self regulating. The idea was to put the foreign fisheries on hold until the size of the stocks could be quantitatively assessed by American scientists.[42]

The Soviet Union caught 27,000 tons of pollock in the eastern Bering Sea in 1969, and then turned their fleet in that direction, catching 219,000 tons two years later. The peak in Soviet catches occurred in 1973 at 280,000 tons. The Soviets were mainly freezing the fish for domestic consumption. South Korea and Poland were relatively minor players in terms of catches, with peaks of 85,000 tons for South Korea in 1973 and 20,000 tons for Poland in 1980.

Although the tools were available, there literally was no management of the pollock fisheries off the coast of North America until the Fishery Conservation and Management Act (FCMA) of 1976. Before that, the pollock harvesting was conducted by foreign fisheries, and even though catch quotas were imposed, the rules had no teeth. Clem Tillion, a prominent figure in the pollock industry, said, "The State Department saw fisheries were of so little value to the United States, that it could be used as a tool to negotiate with other countries."[43]

However, by the late 1970s, the foreign fisheries in the eastern Bering Sea were overbuilt and heading toward a disaster. The situation was even more exaggerated in the Gulf of Alaska, where the population sizes were smaller but the fishing pressure was intense. Over the years from 1964 to 1976, there were 14 million tons of pollock pulled from the eastern Bering Sea. Catches were widely believed to be underreported. This was all about to change in 1976 with passage of the FCMA.

[1.] The *Keiko Maru*, a factory crab processing ship that hosted the author's first voyage in the eastern Bering Sea. A catcher boat is seen making a delivery at midship (from FMA/AFSC).
[2.] An adult Alaska pollock on the deck (from J. Orr, RACE/AFSC).

[3.] A catch of Alaska pollock (from RACE/AFSC).

[4.] A codend of pollock trailing behind the ship. This would be similar to the codend that Røkke scrambled out on to sew up the ripped netting (from FMA/AFSC).

[5.] A trawl net full of Alaska pollock from the Japanese foreign fishery in the late 1970s. Pictured is the codend where the fish are collected in the end of the trawl (from FMA/AFSC).

[6.] Codend of pollock being dumped in the fish bin (FMA/AFSC).

[7.] Observer Wayne Palsson standing in processing bin of Alaska pollock. The trawl is emptied into the bin; the fish then travel on a conveyor belt into the factory for processing (from FMA/AFSC).

[8.] Below deck in the factory on a Japanese factory ship (FMA/AFSC).

# 4 Americanization! The Rush for White Gold and the Developing Fishery

In Korean folktales there is a spirit called a *Dokkabei*, which is often portrayed as a red devil, full of mischief and fond of practical jokes. One afternoon, Dave Stanchfield, the owner and skipper of the *Morning Star*, was angry as heck. He put on a red survival suit, jumped into the icy waters of the Bering Sea, and grabbed on to a trawl net full of pollock to ride it through the chilly sea as it was delivered to a Korean mothership. The net was surging in the swell and he'd occasionally go under, flailing with one arm to stay on the surface while hanging on for dear life with the other. When the big bag of fish was hauled up the stern ramp of the big ship it almost rolled over on him. He had to scramble to keep from being crushed by seventy tons of pollock.

Dave was somewhat of a legend in Alaskan fisheries.[2] He was a "tall, burly man with a flaming-red beard and volcanic eyes," a big voice and a fringing mane of red hair. He had a hot temper, and when he got mad his face would flush from the bottom to the top like a rising thermometer. He must have made quite an entrance as he rode up the stern ramp cursing in his booming voice at the Koreans all the way. The startled Korean sailors probably thought that a red-haired and crimson-suited Dokkabei had been hauled up from the depths of the sea.

Stanchfield was a crab fisherman who joined the pollock trawl fishery as the crab stocks declined. He was involved in a new enterprise in the US fishery: catching groundfish in the US Fishery Conservation

Zone (FCZ) and delivering them to foreign factory processing ships. Stanchfield was in a joint venture (JV) delivering pollock to a Korean processor. The Koreans were commonly known to shortchange the American fishermen. Stanchfield thought he was being cheated and they were underestimating the weight of the nets of pollock he had delivered, so he decided to pay a visit to the Koreans and sort it out. Apparently he did just that.

Stanchfield wasn't the only one to adopt this daredevil method of vessel transfer. There are stories about Kjell Inge Røkke transferring from one vessel to another when the seas were too rough to launch a skiff by tying himself off to a buoy and jumping in with it. The receiving ship would cruise by and snag the line between Røkke and the buoy with a gaffing hook and haul him aboard. It seems absurd to hurl yourself into the Bering Sea when the seas are too rough for a boat, but fishermen think differently from scientists. Tor Tollessen said that he would hitch a ride on a codend as it passed from ship to ship or jump into the sea with a buoy in order to change ships and make electronic repairs.[3] He wore a dry suit and put all his electronic testing devices inside with him, but the suit wouldn't zip all the way up due to the bulk of the devices and he'd get a little damp. The fishermen would assess the risk, but generally would take it without hesitating.

Characters like Stanchfield were the pioneers of the new West Coast groundfish industry. The new joint venture was a daring business and it attracted a group of bold fishermen and investors, like Stanchfield, who were capable of taking physical and economic risks. They were strong individuals who wouldn't be pushed around. Many of the joint venture fishermen became owners of large fishing enterprises, and legends grew around them. Other times, the risk didn't pay off. In 1987 Stanchfield himself took an economic risk in rebuilding an oil supply ship into a factory trawler, the 204-foot *Aleutian Speedwell* (named after the ship that sailed with the *Mayflower* delivering Pilgrims to the New World). Things didn't fare well in that venture, as the *Speedwell* went bankrupt.[4] Stanchfield took a big hit, and eventually he left the fishing industry altogether.

In the 1970s American fishermen like Stanchfield began to take notice of the enormous amount of groundfish the Soviets and Japanese were removing from the eastern Bering Sea shelf. They began to eye riches of the waters beyond the territorial limits of the US boundary, and to worry about the destructive "slash and burn" fishing practices of foreign fishing companies plying the coastal waters. Everything changed with the passage of the Fishery Conservation and Management Act (FCMA) in 1976, whereby the United States unilaterally extended the FCZ boundary out 200 miles into the sea. Although the federal government would still permit foreign fishing, the priority for quotas was granted to US fishermen.

Alaskan fishermen strongly supported the 200-mile limit, but there was considerable opposition to the FCMA in some quarters, especially the shipping industry, the tuna fishing industry, the State Department, and the military. They were concerned about restricted shipping lanes and restricted access to patrol international waters. After passage, Clem Tillion, the "fisheries czar" of Alaska during Governor Wally Hickel's administration, remarked in an interview that the US fishermen had partnered with the oil industry to pass the FCMA.[5] The oil industry was the big winner because they got offshore leases extending out to 200 miles, preserving their ownership of oil wells in the Gulf of Mexico.[6] Supporters of the legislation included most state governments, and West Coast and New England fishermen.[7]

## Joint Ventures

When the FCMA was signed, US fishermen inherited the world's largest fishery. Now they needed to learn not only how to catch pollock, but also how to process and market them.[8] This was going to be a big leap into the ocean, one involving considerable risk. But optimism permeated the domestic fishing industry. They expected a quick end to the foreign fishery, easy access to Asian markets, and rapid growth. It didn't happen exactly as planned. Gaining access to the Japanese and other international markets would be a major difficulty.

A few people immediately recognized the opportunity presented right in front of them. In 1977, Dr. Wally Pereyra had a comfortable job. He was the Director of the Resource Assessment and Conservation Engineering Division of NMFS's Northwest and Alaska Fisheries Center when he got a call from Jim Talbot inviting him to join a new company called Marine Resources Company (MRC).[9] Talbot owned Bellingham Cold Storage and had been involved in the fishing industry for many years. With passage of the FCMA, he had the idea of a joint venture fishery, whereby US fishermen would catch the fish and deliver them to foreign factory ships for processing and marketing.

The United States lacked both the capacity and the market for such large volumes of fish available off the coasts of Alaska and the Pacific states. Opening the markets of the rest of the world to American-caught fish was going to be a big problem. Talbot became intrigued with the large foreign fleet fishing off the West Coast. Perhaps a business arrangement could be made between his cold storage facility, the Soviet fleet, and the US market? So he wrote a letter proposing the idea to the USSR Ministry of Fisheries (Sovrybflot). They didn't answer, and he almost forgot about the letter. Then after nearly a year, the Soviets got back to him and said they were very interested.

Pereyra came up with the details to implement the joint venture fishery. He thought that American fishermen could catch the fish and take advantage of the existing marketing devices. Pereyra recognized the joint venture industry as both a big risk and an incredible opportunity. He had a PhD in fisheries science from the University of Washington and a stable job, which he described as the "best job anyone could hope for." He also was the breadwinner of a family with five mouths to feed. On the other hand, his father, whose approval he sought as a youth, was a Wall Street stockbroker and junior partner at Kidder, Peabody & Co, investment house. Given this business background in his family, his father must have had private thoughts about Wally's chosen career as a fisheries biologist ("you want to be a what?"), but he was supportive.

Today, Pereyra is a busy man, but he was willing to meet with me and be interviewed. Over the next six months we talked four times and exchanged numerous letters. My first contact with Pereyra was in January 2011. He was at his home in Sun Valley, Idaho, skiing. We arranged to connect by a video conference call after his day on the slopes. Pereyra is in his seventies, with graying hair, tanned skin, and a healthy look. He is the founder and Chairman of Arctic Storm Management Group. Everyone I talked to, from the environmental community to industry and government, holds him in high regard.

When we started our video interview there was still daylight in his house, and the room lights were off. His eyes sparkled with enthusiasm. He talked fast and fluidly. As we continued our conversation into the evening, the room darkened as the sun set. His face gradually became front-lit from his computer screen and he became a glowing dis-embodied head floating in a black background. I thought for a moment that I was talking to an apparition from the spirit world or maybe the Cheshire cat from *Alice in Wonderland*.

Pereyra's parents nurtured a strong work and entrepreneurial ethic in their son. During the Great Depression his father's job survived the crash on Wall Street, but life was tough. The family had a small prop-erty in New Jersey and his mother raised chickens. Some mornings she would package two or three dozen surplus eggs in a nice present-able box. The elder Mr. Pereyra, dressed in a banker's suit and hat, car-ried the eggs with his briefcase on the train into the city and sold them to his office colleagues. As a youth, Wally himself tested the waters of small business by trapping and selling muskrats, growing Angora rabbits for pelts, and even forming a small herbal tea enterprise. There was a family history of entrepreneurship and an internal drive push-ing him to make the leap into industry.

From the American standpoint, the joint ventures had the possi-bility to be lucrative. Even though hake and pollock were selling at pennies on the pound, they were caught in large volume. A single tow could net tens of thousands of pounds. The hake fishery alone was catching around 100,000 tons each year. The motive of the Soviets to enter the joint venture wasn't quite known, but one suspects that they could see competition with the Japanese for American groundfish, and they wanted to maintain their supply of fish, as well as their presence off the US coast. There were numerous reports that the Soviet fishing boats were really spying on naval operations and cities along the West Coast.[10]

Getting the joint venture started was tough and a lot of operational details had to be worked out. The first joint venture fishery was for Pa-cific hake. The West Coast fishermen had operated bottom trawls, but flying a large midwater trawl was another experience. In their first attempt at trawling, MRC engaged a fishing boat too small to move the trawl through the water. They discovered that they had to transfer the operation to a bigger ship. After they learned how to catch fish, they also had to work out how to get the codend transferred from the catcher vessel to the Soviet processor. In the first attempt at a transfer, the operation was botched and the net, along with its fish, sank to the bottom. Not only was this an expensive loss, but also it was embarrassing to the US fishermen as the more experienced Soviets stood by and watched. Pereyra and his fishermen had to work out a new system of transferring the codend full of fish from the catcher boat to the Soviet processor.

The late 1970s was the height of the Cold War between the United States and Soviet Union, and they couldn't exchange cash because of the tense relations. Additionally, Sovrybflot did not want cash because they would then have to deal with the Soviet Ministry of Foreign Trade and the Bank for Foreign Trade (Vneshtorgbank).[11] Dealing with the bureaucracy would be cumbersome and the organizations would require a significant "fee." So they developed a barter system. At first they traded hake for Russian king crab, which would be resold in the United States or abroad. Later they also bartered for pink salmon, herring roe, and pollock that were sold on the Japanese market. Other bartered products, such as frozen Atka mackerel and yellowfin sole, were sold in the Egyptian and west African markets. Each trade product was tied to the cost of one ton of hake delivered to a Soviet processor, or a "hake unit." A key to the success of this bartering system was the surreptitious discovery by one of Pereyra's US representatives on the Soviet factory ship of a trading-value key that the Soviets used to determine the value of each species. Although Pacific hake is not eaten by US consumers because it turns mushy after capture, for some reason the Soviet bureaucracy had rated it as highly desirable, on par with cod. This allowed the US fishermen to trade hake for some higher value species.[12]

The barter system sometimes proved to be a risky business. The MRC representatives were exploring new territory. One never knew what was going to happen. Pereyra told one story about how they had arranged to barter hake for 3,000 tons of frozen block pollock. They had lined up a buyer in Mrs. Paul's, which had contracted with a processing plant in South Korea to make the frozen pollock into fish sticks. Mrs. Paul's inspected a sample of headed, gutted and tail-less product provided by the Soviets and negotiated a price based on estimated yields. But when the fish were delivered they still had the tails. Mrs. Paul's refused to pay for the product because their yield calculations were based on tail-less fish. Eventually they settled the dispute. Then they had to deal with several ministries in the USSR as well as the Korean CIA to get the product off the Soviet ships behind the Iron Curtain and into the South Korean–based factory.

In the early 1980s, Pereyra and several colleagues split from MRC over a dispute with the Soviets and formed their own company, called ProFish. ProFish got involved in the pollock joint venture, catching pollock and supplying it to processing ships from five or six Korean companies.

Pereyra told another story about buying the ship *Smaragd* to fish for pollock in the Bering Sea. He took a second mortgage out on his house to have the ship rebuilt from an east coast scallop boat into a pollock trawler.

Initially, the rebuilt ship failed the US Coast Guard's incline test for stability, and they had to move some heavy winches beneath decks for it to pass inspection. Then after the ship left Seattle in February for the Bering Sea, he got a call from the US Coast Guard in the middle of the night to tell him that the ship was in heavy seas in the Gulf of Alaska and sinking. The crew was standing by to abandon ship. As a last resort, the helicopter hovering over the ship to rescue the crew lowered a pump they had aboard to see if that would save her. Pereyra said he saw his whole life flashing in front of him. Over the next few hours of waiting he thought he would lose everything, and he feared that the lives of his crew were in danger. As luck would have it, the pump worked and the ship was saved. The *Smaragd* made good money that season.

Pereyra was an educated and knowledgeable scientist who made the leap into industry. He's been described as a "go getter and a hard worker."[13] He was unusual in that he entered the business world from the gates of science rather than from the fishing deck. Like the others who followed him into the fledgling industry, he took huge risks. With his background in science and industry, Pereyra served for nine years on the North Pacific Fishery Management Council (NPFMC), and he is probably one of the main reasons that the Alaska trawl fishery got off to a good start as a well-managed fishery.

One of the fishermen working with Pereyra in the joint ventures was a "larger than life" character named Barry Fisher.[14] Fisher was born in Gloucester, Massachusetts, and started fishing at age twelve as a "catchy" on a longline dory. He dropped out of school after eighth grade and eventually began running his own boats.[15] After serving in the US Army, Fisher completed his General Equivalency Degree (GED) and went on to graduate from Harvard University.

After Harvard, Fisher tried life in corporate America, but he realized it wasn't for him, saying, "I really believed these corporations wanted creativity and innovation from me, when what they really wanted was my soul."[16] He went back to fishing. Later he was offered a position as associate professor at Oregon State University to teach about fishing technology, and he moved to the West Coast.

But Fisher couldn't keep away from the action on boats, and after five years of teaching, he resigned and started trawling again. In 1978, Fisher joined Pereyra in the joint venture operation for hake, and later joined the pollock fishery in Shelikof Strait. Fisher was demanding of his crew. Dave Fraser, later to become owner and skipper of the *Muir Millach*, was a deckhand for Fisher in those days, and tells a story about reclining in his bunk after a hard day of fishing. He looked at the ceiling above him where crew who'd slept there before him that season signed their names. Dave's name was number thirty-nine on the list.[17]

Fisher was smart, outspoken, and opinionated. Exclaiming the independent nature of his breed, he said, "Fishermen are not domesticated. They're really hunters and the best hunters know that there are lush times and lean times." He continued, "Managers are not hunters—they depend on salaries and place the greatest value on security. They continually try to mutate the fishing industry into something even and controlled. That may be a laudable goal, but it's just not very realistic and it usually turns out to be a waste of time."[18] I would go a step further and say that the qualities that make good fishermen and good managers are completely opposite. Acceptance of risk and boldness in a fisherman are replaced by uncertainty and caution in a manager. A manager who exudes unrealistic confidence about his numbers is probably one to be wary of.

The joint ventures were slowly developing, but a catastrophic event occurred in the Bering Sea that was to prod American fishermen to participate in the pollock fishery with the Soviets, Koreans, and Japanese. In 1980, about 50% of the crab fishing industry was owned by Norwegian immigrants.[19] Around that time the crab stocks in the Bering Sea took a steep downward turn. Whether the crash of the crab

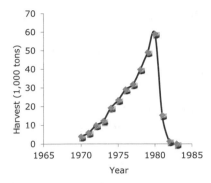

Figure 4.1. The build-up and collapse of the US red king crab fishery in the eastern Bering Sea. After the collapse the crab fishermen were desperate. Many changed to fish for cod and pollock.

populations was related to crab bycatch in the groundfish trawl fisheries is still controversial, but it would be an ironic although not too surprising twist in the story.

## Fish and Chips

According to John Sjong, when the crab stocks started to decline, "we started looking around: what else can we do?" Another prominent Norwegian-American, Konrad Uri, said that the riches of crab "brought a lot of new people into the industry, doctors, lawyers, you name it." As a result there were too many people fishing for a limited resource, causing Sjong to remark: "There was a tremendous number of boats brought into the industry, there was no way it would hold up, the writing was on the wall." With stunning declines in Bering Sea crab stocks in the early 1980s, a large portion of the Norwegian-American fleet began converting to groundfish trawlers, many of which hooked up with the "joint venture" fishery.[20]

One group of prominent fishermen collaborated to look for opportunities in other fisheries and formed a group, called the Highliners Association, in 1980.[21] They were mainly crab fishermen who were concerned about what they would be catching in the future. The Highliners became known as the "Seattle Mafia." The group contracted with Natural Resources Consultants (NRC) to evaluate, first, the potential of the pollock fishery as an alternative to fishing crabs, and second, what could be done to hasten "Americanization" of the pollock fishery. The NRC was founded by Dayton Alverson, former director of the Northwest and Alaska Fisheries Center, and included the renowned economist James Crutchfield on its staff. When the NRC recognized that overall costs were impeding progress, it proposed to examine the FCMA to see if there were some provisions to assist in the industry's growth. They proposed a meeting with the Japanese to convince them of the need to develop the joint venture.[22]

The FCMA included a three-tiered quota amendment by which the total allowable catch was prioritized among categories of catchers. The highest priority was given to US vessels catching the fish and delivering it to US processors. The second tier was US vessels catching and delivering to foreign processors, and the lowest tier was the leftover going to foreign catchers and processors.

Because the US pollock fishing industry was having a hard time taking flight, NRC proposed a "stick and carrot" approach to encourage the Japanese to buy fish from the joint venture vessels. Under this scheme an amendment was proposed that would give priority access to those foreign fisheries that helped the US fisheries with joint ventures or helped develop shoreside processing facilities.

This was the so-called Fish and Chips Amendment that was passed by Congress in 1980. It took a while to engage. A meeting was held in Seattle in 1982 and Japan agreed to buy a quota of about 200,000 tons of pollock from US catcher vessels. About a half-dozen catcher boats from the US crab fleet were needed to do the job. By 1985, the joint venture fisheries had taken off and two new shoreside surimi plants had been built in Alaska.

The joint ventures, especially for pollock, developed rapidly into a success and it was lucrative business for the fishermen involved in it. But the process of total "Americanization" was moving too slowly for some people in the industry. Why should the Americans be catching the fish only and missing out on the added value of processing it?

### Americanize!

In 1985 the Pacific Seafood Processors Association was awarded a competitive grant through the US government's Saltonstall-Kennedy program to investigate what could be done to completely Americanize the fishery and to do it more quickly. Through a subcontract, a report was produced for the processors association by Alverson's NRC with specific recommendations on implementing Americanization.[23]

This report outlined the problem of the stuttering progress of "Americanization" and a plan for action. According to the report, the problem arose from the higher capital, labor, and insurance rates faced by the US industry compared with the foreign fishery, putting them at a cost disadvantage. The blame was put on US policies such as the Jones Act and various safety and environmental restrictions. "US harvesters have to pay more for their vessels on a comparable basis; largely because of national economic and maritime policies, [this] translates into a heavy financial burden for the average fisherman."[24] In this capital-intensive industry, they calculated that foreign fisheries had more than eight times more fixed assets per worker already in place compared with the US industry. Investments would have to be made in order to catch up. They further argued that other nations were subsidizing their fisheries. In addition, labor wages of US workers were much greater than those of the foreign fisheries; for example, the Koreans were paying about fifty cents per hour for fish processing workers.

The NRC document went on to report that in order to improve markets, they advocated trade barriers and high tariffs, and development of protected specialized markets such as those provided in programs of the Defense Department (US Army meals) and the US Department of Agriculture (school lunch programs).

The treatise also directed that policies should be implemented for low-interest loans, direct subsidies for vessel construction, and protection of markets from foreign competition. Costs related to safety, sanitation, and manning requirements should be minimized. In other words, the US government could be blamed for the US fishery being noncompetitive. Subsidies would be required to make up for the differences, and regulatory laws would need to be relaxed to keep the industry profitable.

The report further suggested that allowing the joint ventures to operate without limits would hinder the development of at-sea and shoreside processing capability. "Strategic allocation of resources surplus to US needs may be the single most important tool influencing the continued process of Americanization."[25] The Japanese resisted giving US fishermen access to technology, and the Japanese threatened to switch to Chilean mackerel for surimi because, they said, "the supplies of Alaska pollock were unreliable."[26]

As the US industry developed the surimi products acceptable to the Japanese market and along with the growing catch and processing capabilities, the US fishery displaced the joint ventures through the three-tiered amendment of the FCMA. The plan of the Pacific Seafood Processors was enacted through the political process.

Foreign fleets had paid an access fee to fish in US waters. The rates were not negotiated but all vessels paid the same rate. There were four types of fee: (1) vessel permit fee, (2) poundage fee based on the catch, (3) surcharge fee, and (4) observer fee. In 1982 the poundage fee on foreign fishing vessels brought in $34 million.[27] However, the fact that US vessels did not pay for fishing rights set an important precedent, which is the fish are a public and common access resource, free for those with the means of accessing it.

Wally Pereyra and others in the joint venture operations could see that Americanization of the fishery and the three-tiered priority for fish quotas established by the FCMA was squeezing out the foreigners, first from catching fish and then from processing them. The joint venture fishery would be short-lived. In 1984 ProFish negotiated a partnership with a group of investors, including Oyang Fisheries of Korea, to convert the retired Navy tanker *Patapsco* into a surimi factory trawler named the *Arctic Storm*. The *Arctic Storm* was financed by Christiana Bank and designed in collaboration with Jensen Maritime Consultants and Fiskerstrand & Eldøy of Norway. It was rebuilt in the Seattle shipyards of Wright Schuchart at a cost of $25 million.[28]

In 1986, Arctic Storm Inc. began. As of this writing, Pereyra continues as the founding major partner and chairman of the Arctic Storm

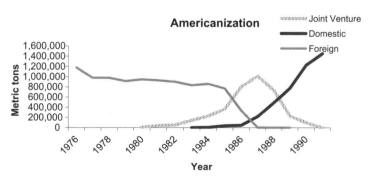

Figure 4.2. The Americanization of the fishery showing the shift from foreign to joint venture and domestic catches of pollock.

Management Group, which remains a major player in the pollock industry.

Pereyra reasoned that there are three types of entrepreneurs in the business: pioneers, adapters, and followers. Of those, the adapters are the most successful. Pereyra says he adapted the Japanese concept of catching and processing fish at sea as his business model. Norwegians entering the fishery, such as Sjong and Røkke, had a similar concept based on Norwegian factory trawlers.

Norwegian investors began to finance a massive rebuilding of US ships into trawlers, factory trawlers, and processors, with much of the work done in Norway and financed by Norwegian interests.[29] There was a huge amount of money at stake as the typical cost of converting a bare-bones hull into a large catcher processor was $38 million.[30] Norway's Christiana Bank alone had a loan portfolio in the Seattle-based fishery of $435 million, and other Norwegian banks such as Creditbank and Bergen Bank also had large loan portfolios.[31] There were also US and Japanese subsidies to rebuild and modernize the fleet. Pereyra said, "you could get into a factory trawler with nothing down."[32]

Whereas the Norwegian banks were heavily invested in the offshore processing of US-caught pollock, the Japanese invested in shoreside processing with catcher vessels delivering fish. The ensuing struggle for control of the resource could be viewed as a war between Norwegian and Japanese interests over American fish. By 1989, Japanese interests controlled 85% of the shoreside processing capacity. The onshore and offshore fisheries sectors clashed over allocations.

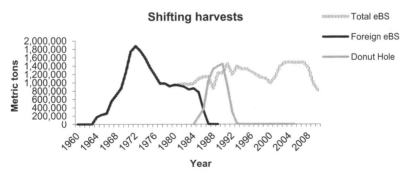

Figure 4.3. Shifting of the harvests in the eastern and central Bering Sea, showing that as the US fishery increased, the foreign fishery decreased and shifted into the "Donut Hole." eBS indicates eastern Bering Sea.

The result of the build-up in capacity of the US fleet was the "Americanization" of the fishery. Ultimately, the ability of the American fleet to catch and process the total harvest quota for pollock in US controlled waters forced the foreigners out of the direct fishery.

But sometimes well-intended actions have unforeseen consequences, leading to new problems. The three-tiered system of the FCMA mandated that Americans had first priority to the fish inside the 200-mile limit. When the American fishing capacity increased to the point of being capable to catch all the fish, they took over the fishery in the eastern Bering Sea. The foreigners had to pack up and leave. The foreigners also had to keep fishing to pay loans and to keep factories running, people working, and shelves stocked. Where did they go? Some of the ships started fishing for pollock in the western Bering Sea and Sea of Okhotsk, but many ended up fishing in the "Donut Hole."

# 5 An Empty Donut Hole: The Great Collapse of a North Pacific Pollock Stock

*Would they have enough if all the fish in the sea were caught for them?*

Numbers 11:21–23

It was a blustery gray day in February 1986, and I was on a NOAA research ship in the middle of the fishing fleet in the "Donut Hole," the international zone in the middle of the Bering Sea nested between the territorial waters of the United States and those of the USSR. I counted sixty large factory trawlers around us belonging to five nations. They lined up in a pattern of several rows to take turns dragging across a thin layer of Alaska pollock at about 400 m depth, fishing with cavernous nets that opened forty-five meters vertically for durations of several hours. That year the Donut Hole sustained a "reported" winter catch of about one million tons. In hindsight, I was witnessing the extirpation of pollock in the central Bering Sea.

The Donut Hole is in the abyssal waters of the central Bering Sea beyond the continental shelf. The Donut Hole is like several other international "no man's lands" in the world's oceans. The "Smutthullet" in the Barents Sea, which translates from Norwegian as the Loophole, lies in the international waters between Norway and Russia, and the "Peanut Hole" in the Sea of Okhotsk is an international zone encompassed entirely by Russia. These are troubled spots in our oceans, surrounded in controversy about fisheries piracy, illegal fishing, and overfishing.

The little-known rise and fall of the pollock fishery in the Donut Hole and the surrounding waters of the central Bering Sea during the 1980s is one of the most spectacular collapses due to overfishing to occur in the modern history of world fisheries. It shares the dubious

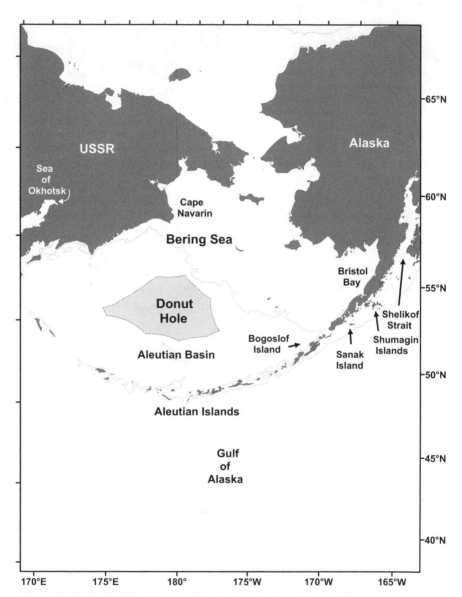

Figure 5.1. The Donut Hole of the central Bering Sea. The Donut Hole is in the international waters beyond the exclusive economic zone (200-mile limit) of the USSR and the United States.

honor of the largest fishery collapse in the Northern Hemisphere with the coastal Norwegian spring-spawning herring collapse of the 1970s. If nothing else, it is a controversial sequence of events that had tragic consequences. How did this happen and escape notice?

The collapse of a fishery is sometimes defined as a decline in catches to 10% of the maximum previous level.[1] Collapses of fisheries usually cause tremendous economic and social problems. They also cause a lot of exasperated finger-pointing and laying of blame among scientists and politicians.

Monastic and tax records surviving from the tenth century document fluctuations in the Bohuslän herring fishery near the southwestern coast of Sweden. These old historical records show that herring would be abundant for periods of 20–50 years and then disappear for 50–70 years. The ebb and flow of the herring and irregular patterns of abundance over the past millennium caused periods of feast and famine in the local economy.[2] Herring were the main trading commodity of the Hanseatic League in the thirteenth and fourteenth centuries, and the herring along the southern coast of Sweden were convenient to salt mines near Kiel, Germany, for preserving the fish. The periods of collapse caused havoc and great hardship in the local economy.

Periods of boom and bust are common in marine fisheries. In 1945 catch records for the California sardine were broken; 9,000 tons were delivered to the port of Monterrey in just a single day. The next year a population crash occurred that was catastrophic.[3] Prices for the fish skyrocketed and fisheries industries, such as the Van Camp and Starkist companies, scrambled for other sources of raw material for their reduction fishery to animal feed. They both ended up fishing in Peru for the anchoveta. The historian Arthur McEvoy describes the population crash as "the collapse of the California sardine fishery, which at the time was probably the most intensive fishery the world had ever seen, and the loss of which ranks as one of the most egregious failures in the history of US wildlife management."[4]

Over time, the Peruvian anchoveta fishery also collapsed under the dual stresses of environmental change and fishing pressure in the 1970s. It was probably the world's largest fishery collapse ever. The economic and ecosystem stresses caused by the fall were felt around the world.[5]

The collapse of the northern cod population off the coasts of the

maritime provinces of Canada in the 1980s is one of the best docu-
mented modern fisheries collapses. It was almost certainly caused by
overfishing. Finally, a moratorium on fishing cod put 40,000 people
out of work in five provinces, an event sometimes called the biggest
layoff in history. Newfoundland lost 10% of its population in the decade following, and compensation programs cost $3 billion.[6]

Rapid declines in fish populations also cause major ecological problems. Disappearing fisheries generate trophic impacts at both lower
and higher levels of the food web, causing shifts in communities and
highly visible impacts on seabirds and marine mammals that have
grown to depend on the resource.[7] Sudden massive die-offs of seabirds have been linked to changes in the distribution and abundance
of their fish prey, and decreased levels of pollock have been linked to
declining numbers of Steller sea lions and northern fur seals.[8]

Given what was known about the life history of pollock as continental shelf–dwelling fish, it came as a surprise in the 1970s when a
large population was found in the very deep water over the Aleutian
Basin. A fishery for pollock quickly developed and fisheries scientists were caught off guard. Almost all the pollock caught here in the
1980s were netted exclusively by non-US fisheries, mainly those of Japan and the [former] USSR, but also China, Poland, South Korea, and
others. Pollock in the Donut Hole were believed to be a "straddling"
population, or part of a larger population in the Aleutian Basin that
moves across international boundaries, but no one really knew where
the fish came from.

It is a difficult problem to assess how many fish once existed in the
Aleutian Basin, but fishery population models estimate that nearly
13 million tons of pollock age five and older were in the US zone of the
Aleutian Basin in 1983.

This number was calculated from a model that included reported
catches from the Donut Hole region and from the fishery in US waters in the Aleutian Basin near Bogoslof Island.[9] The highest officially
reported catch in the Aleutian Basin, including US and international
Donut Hole waters, but excluding Russian territorial waters, was
1.7 million tons in 1987. By far the greatest harvest area was the Donut
Hole. By 1988 the biomass plunged to less than 50% of its peak (while
the catches were still increasing) and by 1992 was only 6% of the maximum; by the above definition of declines in harvests, a collapse had

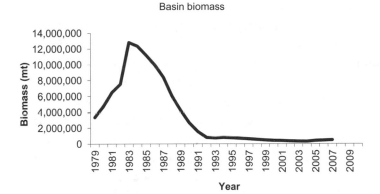

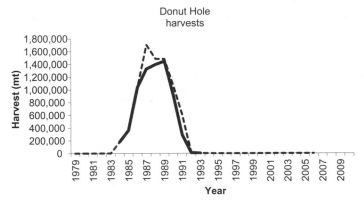

Figure 5.2. The biomass of pollock in metric tons (mt) of age 5 and older fish in the American zone of the central Bering Sea basin, including the Donut Hole and Bogoslof Island stocks (top panel). The lower panel shows reported harvest levels in the Donut Hole (the dashed line includes the Bogoslof Island catches).

occurred, and this is supported as well by the estimates of decreasing biomass. By 2007, the biomass estimate was at its nadir of 309,000 tons, a drop of 98% from the maximum.

One of the uncertainties fisheries scientists have to deal with is whether the population abundance prior to the availability of fisheries statistics increased as shown in Figure 5.2 or whether it was at a high level at some point in the past as well. Was the increase in the stock just a temporary anomaly, or were the fish always abundant there? Part of the problem lies in the models used to assess population abundance because the numbers are calculated from the numbers of fish caught, and it follows that if no fish were caught then there were no fish there. By extension of that general idea, as the fishing effort of a new fishery increases and expands, this has a tendency to show increased biomass as a function of discovery and increasing efficiency. We assume that fish were abundant in the Donut Hole before the fishery; we just don't know exactly how many there were.

The demise of the Aleutian Basin pollock population has received scant attention compared with declines of other fish stocks. A search through the referencing source *Science Citation Index* reveals only six publications with the topics/titles of pollock and Aleutian Basin. Only one publication involves both the terms pollock and Donut Hole, and none document the collapse. By contrast there are 266 publications with the topics of cod and Canada, and 212 publications with the topics of sardine and California. This observation is despite the fact that the Aleutian Basin pollock collapse was over three times larger than that of the sardine or cod collapses.

Whereas the biomass of pollock in the Aleutian Basin went from nearly 13 million tons in 1983 to 309,000 tons in 2007 (Figure 5.2), Pacific sardine had a spawning biomass of about 4 million tons in 1935, which declined over a period of nearly 20 years to about 200,000 tons. Annual catches declined from around 800,000 tons to less than 100,000 tons over the same period and to near zero by 1968. For the northern cod, which comprises a cluster of individual stocks off the coast of eastern Canada, the spawning biomass was 1.6 million tons in 1962 and declined to less than 10% of that value by 1992.[10] Catches peaked at 800,000 tons in 1968, and by 1992 the northern cod fishery was closed.

## A Developing Fishery

Japanese scientists reported finding large concentrations of pollock in the Aleutian Basin in the 1970s and reported finding spawning fish in the Donut Hole in 1983.[11] By then, the scientists were already a step behind the commercial fishery, as a major harvest had already started. "Official" catch records indicate that the fishery in the Aleutian Basin began in 1984, but it is clear from US observer records that significant fishing had been happening since at least 1981.[12] The fishery operated largely in winter to catch prespawning females for their valuable roe.

The remaining carcasses and male fish were processed into "headed and gutted" frozen fish, surimi, and fish meal.

The Donut Hole fishery did not really escalate until Americanization of harvests within the territorial waters of the United States occurred and pushed foreign fisheries into the international zone of the Aleutian Basin, since US fishermen had first priority to the "total allowable catch" in the FCZ out to 200 miles as a result of the FCMA of 1976. When the US fleet reached the capacity to catch all of the total allowable catch within US territorial waters, the foreign fleet was squeezed out of US territorial waters and into the international Donut Hole, and the fishery there intensified. In 1985 the reported catches from the Donut Hole reached 360,000 tons of fish. There were attempts to shut down the Donut Hole fishery. As early as 1988, Alaskan Senator Ted Stevens pushed through a resolution in Congress to call for an international moratorium, but fishing continued unchecked.[13] Reported harvests in the Donut Hole nearly quadrupled to 1.5 million tons in 1989. Foreign scientists reported that the stock was huge, up to 15 million tons, whereas US scientists believed it was much smaller. Two years later, the bottom fell out—the harvest dropped to 293,000 tons.

An international meeting was held to discuss curtailing the fishery. Japan was willing to look at the issue, but other countries resisted. Poland argued that between 30,000 and 50,000 jobs depended on the pollock fishery, which were critical to the developing democracy after the fall of the Iron Curtain. China had just recently purchased fifteen large factory trawlers and they needed to be working, and South Korea pointed out that they had been excluded from the surrounding Exclusive Economic Zones (EEZs) and had nowhere else to go for pollock.[14] By 1992 the fishery was nearly extinct with a take of 10,000 tons. In 1993, a moratorium was put in place by international agreement just before the 1994 signing of the Convention on the Conservation and Management of the Pollock Resources in the Central Bering Sea (hereafter the Central Bering Treaty).

### Other Basin Populations

About 400 km to the southeast of the Donut Hole, Bogoslof Island pokes up through the surface from the abyssal plain of 3,000-meter depth. Catch statistics for the Aleutian Basin include the fishery around Bogoslof Island, but the connection between the Donut Hole and Bogoslof populations, although accepted by many stock assessment scientists back then and even now, lacks substantial and direct evidence to date. The prevailing view of stock assessment scientists is that the fish moved freely between the Donut Hole and Bogoslof Island regions. The fishery on pollock around Bogoslof Island was even more short-lived than that in the Donut Hole, lasting only about five years. A new joint venture fishery in the Bogoslof region started in 1987 with a catch of 340,000 tons; in 1989 the harvest was 39,000 tons, and then in 1992 catches dropped to 241 tons and the fishery was suspended before the signing of the Central Bering Treaty in 1994. The drop in harvest levels is mirrored by a drop in NMFS acoustic survey estimates of biomass from 2.4 million tons in 1988 to 110,000 tons in 2007. However, the decline may be underestimated because the survey area has shifted; since fish became much less abundant over the deep waters of the Aleutian Basin, the survey now covers more nearshore waters than were surveyed in the earlier years. Since then catches have varied from between 10 tons and 100 tons per year.

Pockets of spawning fish have been seen in the area just to the north of the Aleutian Islands (e.g., near the Islands of Four Mountains and Baranof Ridge) by surveys. A fishery developed along the Aleutian chain in 1979 at a level of about 10,000 tons and grew to a peak of 50,000–100,000 tons in the early 1990s. Then, in the late 1990s harvests in the Aleutian Islands dropped to about 20,000 tons, and in 1999 the harvests were severely restricted by the North Pacific Fishery Management Council (NPFMC), ostensibly to protect endangered Steller sea lions. Over that time frame, the peak biomass of age 3 and older pollock in the Aleutian Islands dropped from 1.6 million to 170,000 tons, an 89% decline.

Over on the Russian side of the Bering Sea, it is more difficult to document what has happened. Reported catches in the southwest Bering Sea peaked at 1.13 million tons in 1981 and declined steadily through the 1980s and 1990s to 79 thousand tons in 1995 and only 4,000 tons in 2006, or less that 1% of the maximum catch.[15] Both American-based vessels and those flying the flags of other nations were involved in fishing for pollock in the southwestern Bering Sea. On the other hand, in the extreme northern part of the Bering Sea near Cape Navarin, catches have remained fairly steady.

## Causes of the Decline

What went wrong in the Donut Hole? A cascade of events concentrated a large fishery into a small area, and catches in the international waters of the Donut Hole were unregulated and probably misrepresented. By the time the fishery was officially halted by the Central Bering Treaty in 1994, it was too late. Japanese scientists estimated that there were 148 vessels fishing for pollock in the Donut Hole in 1986–87, and the efficiency of gear and the methods for catching them were constantly improving.[16]

The Aleutian Basin and its living resources have been little studied before and after the collapse of pollock there. A potential cause for the rise and decline of pollock would be an environmental factor. However, there is no obvious relationship of pollock dynamics in the area with temperature changes.[17] There also doesn't appear to be any change in the abundance of zooplankton prey of pollock that could explain the collapse. Although zooplankton in the Aleutian Basin declined from 1993 to 1995, by then the collapse of the pollock stock had already happened.[18]

Managing fisheries on the high seas and in international waters has a long history. The expanding foreign fleets were operating under the long-standing concept of "freedom of the seas" and everything outside the territorial limit was available on a "first come, first served" basis.[19] Regulating the catch levels of high seas fisheries was difficult, as it had been established in 1955 at the International Technical Conference on the Conservation of the Living Resources of the Sea held in Rome that foreign fisheries could be halted only after scientific studies proved that overfishing was occurring.[20]

The situation in the central Bering Sea was similar to other areas. From 1967 until the FCMA of 1976, foreign fisheries in the eastern Bering Sea were managed by bilateral agreements. Catch quotas were placed on eastern Bering Sea pollock in 1973, but they were self policing and self reporting.[21] These policies set the background for unregulated development of the Aleutian Basin fishery.

Anecdotal observations in the 1980s indicate annual harvests as high as 3 million and even 5 million tons.[22] At that time the biomass of fish in the Aleutian Basin and Donut Hole was uncertain. In retrospect, the harvest levels were still increasing in the late 1980s even as the stock was crashing; the harvest ran ahead of knowledge of the status of the resource, which is common in start-up fisheries.[23] It is clear from historical documents that experts in the field, although recognizing the potential for a dangerous situation, did not realize the declining state of the population in 1989 when the harvest peaked.[24]

Underreporting of catches on a massive scale was suspected, and vessels that were supposed to be fishing in the international zone were suspected of transgressing into US waters and fishing illegally there. In 1988 the US Coast Guard reported 94 sightings of foreign vessels fishing in US waters near the Donut Hole area.[25] The US State Department issued several *Notes Verbales* charging evidence of organized industry conspiracies to cheat.[26] In response, the national government of Japan distributed strict punishment to proven offenders, but effectively patrolling the expanse of the distant sea and catching them was difficult.

Bill Aron, the science and research director at the Alaska Fisheries Science Center (AFSC), said that "foreign vessels were cooking their books. US observers on Japanese boats had found that without exception those cruises had perhaps one-fourth the catch . . . of unobserved boats."[27] Congressman Don Young called it "the international thievery of our marine resources."[28] Ted Evans, the executive director of the Alaska Factory Trawlers Association, said, "We simply can't understand how a country as open as Japan could permit this kind of organized crime to go on."[29]

So-called IUU (illegal, unreported, and unregulated) fishing probably occurred with Soviet pollock fisheries in the western Bering Sea side. This has continued to be a serious problem in the Russian Exclusive Economic Zone (EEZ) of the Bering Sea.[30]

A collision of events resulted in overfishing to the point of "commercial extinction."[31] As mentioned earlier, declining crab harvests in the 1980s, a rise in the price of frozen fish, and a favorable currency exchange stimulated the conversion of many crabbers and other ships to trawlers. The FCMA set up conditions for Americanization of the groundfish fishery in US waters, which had the unintended consequence of forcing the large foreign fishery into the Donut Hole, where it was unregulated.[32]

Bill Aron said, "Neither nation [the United States or the USSR] has any authority over the Doughnut Hole [sic]. There are treaties that allow us to assert rights to our native salmon runs, but there is no body of law affecting pollock."[33] The pollock of the central Bering Sea fell prey to the enhanced acoustic technology to find them and the development of lightweight materials that allowed massively large trawl nets to catch them proficiently. The attitude of high seas foreign fisheries at the time contributed to the collapse; the resources of the open seas were considered boundless—everything outside of sovereign waters was available on a first come, first served basis, and under the strategy of "pulse" fishing, harvesters pounded one area and then moved on to more productive ones.[34] There was still a prevalent belief that there existed a surplus of unutilized fish in the ocean.

### Unanswered Questions

In 1986 the Alaska Fisheries Science Center began to research the pollock in the Donut Hole in response to a growing concern about the foreign fishery there. We had an ambitious plan: our survey was to determine where the fish were spawning, to collect eggs and larvae for natural history studies, and to determine where the fish came from using genetics and the chemical composition of their otolith ear stones to track their birth origin based on elemental fingerprints. Knowing where the fish came from would help us determine how the fish population was structured, and how the fishery could best be managed. At the least, we hoped to determine whether the foreign fishery in the Donut Hole might impact the American fishery in the US EEZ.

Locating a densely packed spawning aggregation in the Aleutian Basin, an area spanning over one million km², is like finding the proverbial needle in a haystack. Fish are found by searching along a transect, traveling at about fifteen km/hour, looking underneath the vessel path with an acoustic beam that is several meters wide. Instead, we decided to rely on US observers on foreign fishing vessels to report by radio where they were finding spawning fish. With more than a hundred vessels plying the waters for several months, this considerably enhanced our probability of finding a spawning aggregation. The strategy paid off; we found spawning fish in locations where observers had reported them.

The results of our studies indicated tantalizing but inconclusive differences in Donut Hole fish compared with those collected from adjacent regions. In hindsight, the technology available and the sample sizes supported by meager funding levels did not have the resolving power to distinguish the small differences we might have expected. The failure to find conclusive results and the lack of funding from NMFS to support further research in this subject was a great professional and personal disappointment.

After 1986, research on the Donut Hole pollock waned partly because the core of scientists interested in the scientific problems had dissipated: I was transferred to another program in the Gulf of Alaska and several others left for academic or administrative positions. Eventually, alarm about the status of the fishery generated efforts to mount a new research program, but a large ship was needed to work in the Bering Sea at considerable expense. When the funding and ship time were finally secured in 1993, the Aleutian Basin population had already crashed. Scientists in this program spent two research seasons looking for pollock eggs and larvae in the area; and when they were unable to find sufficient numbers to justify continuing, they adjusted their objectives and transferred their operations to the eastern Bering Sea shelf, where fish were still plentiful. Since that time, research in the Aleutian Basin by US scientists has been limited mostly to acoustic assessment surveys near the Aleutian Islands.

The question of overfishing was complicated by scientific uncertainty about the connection of the Donut Hole aggregation to other spawning groups in the Aleutian Basin. The issue of the relationship of Donut Hole fish to other stocks remains unanswered after some twenty years. Historically, the Soviets believed that the Aleutian Basin fish were closely connected to stocks over their own continental shelf, and the Americans made a similar claim for themselves—that is, these were straddling stocks; while the countries without a coastline on the Bering Sea, such as Japan, claimed that the Aleutian pollock were an independent stock.[35] As usual, each nation viewed the data in light of its own self-interest.[36]

The consensus view reported in the recent AFSC stock assessment documents is that pollock in the Aleutian Basin were surplus fish overflowing from the shelf population as a density-dependent response. The pollock stock assessment scientist Vidar Wespestad in 1993 stated, "Research shows that pollock harvested in the central Bering Sea move from the adjacent shelf populations. Recruits to the basin are coming from another area, most likely the surrounding shelves."[37] Likewise, other stock assessment scientists have recently written, "Strong year classes of pollock in Bogoslof may be functionally related to abundance on the shelf" and "Collectively, pollock found in the Donut Hole and in the Bogoslof region are considered a single stock, the Aleutian Basin stock."[38]

The main support behind these assertions is that some of the same strong year classes that appeared in the Aleutian Basin were also strong over the shelf. This observation led to the hypothesis of density-dependent overflow into the basin from the adjacent shelf, known as the "serendipity hypothesis," which postulates that fish were from a couple of very strong year classes over the surrounding continental shelf waters.[39] Other biologists, including myself, are less certain about the biological relationship of shelf and Basin, and of Donut Hole and Bogoslof Island spawning concentrations of pollock.

In the early 1900s, the iconic Norwegian scientist Johan Hjort showed that fish stocks were fluctuating not because of large-scale migrations, but because of interannual variability in year-class strength.[40] Although environmental conditions may cause temporary distribution shifts, large-scale and semipermanent displacements of fish populations similar to that described for pollock in the Aleutian Basin are somewhat rare. A well-known shift in Iceland-Greenland juvenile cod is caused by anomalous currents that transport them toward Greenland and away from their usual nurseries; however, fish return to natal spawning grounds as adults.[41] Density-driven migration excursions are also observed in capelin, although the maintenance of these patterns is not clear.[42]

One of the best known examples of large-scale migration is that of the "Bohuslän herring," which appear in great quantities off the west coast of Sweden at a frequency of about once a century and maintain high levels of abundance for several decades. One hypothesis put forth to explain the Bohuslän herring phenomenon is that a very strong year class of herring in the North Sea, combined with an environmental window, allows them to enter deep into the Skagerak, which starts the overwintering pattern. They learn this migratory path, and then other new year classes learn the behavior by following the older fish.[43] However, the respected fisheries biologist David Cushing also suggests that the region could be seeing the effect of waxing and waning local populations.[44] Since the Bohuslän herring has not re-appeared for the last 100 years, direct testing of the alternative hypotheses has not been possible.

The situation in the Aleutian Basin discussed above is a little different from the other fisheries. First, fish were not just overwintering in the Basin, they were spawning there.[45] Second, juveniles were rarely caught in the Basin (only adult fish were caught), leading to the assertion that they would spend their juvenile life on the shelf, and then migrate back to the Basin where they would remain throughout their adult life. Data directly supporting either the notion that Aleutian Basin fish are a self-sustaining population, or that their occurrence is due to an overflow of fish from the shelf to the basin, are lacking, although very much needed. It seems noteworthy that the maximum population in the Aleutian Basin exceeded that ever observed over the eastern Bering Sea shelf, and it seems unusual that a "surplus" would be larger than the source. And such massive displacement over the distance of nearly 500 kilometers that separates the major spawning areas would be unprecedented. The converse may be just as plausible: the basin may have been the source population that seeded the shelf with juveniles.

Another concept, that of "local populations," asserts that pollock, like many other fish populations, may be organized into multiple spawning groups, or stocks, that over the millennia have become well adapted to their chosen habitat. These species include salmon and Atlantic cod, which have discrete spawning populations that may mix during the feeding season but that reassemble and segregate during spawning through homing migrations. This is the perspective of "biocomplexity," which suggests that fish populations may have a "portfolio" of subpopulations that dampen the variability observed in individual stocks.[46] Under these circumstances, as an individual spawning population is decimated or as its habitat becomes unsuitable for spawning, the population may be lost for a considerable time until recolonization establishes a new population. Nature's experiments with colonization can result in a successful self-sustaining population, or in a population sinkhole that attracts new recruits to a poor survival habitat, otherwise known as a "spawning trap."

This idea was supported by a group of Norwegian scientists who wrote that the repopulation of a spawning stock may depend on adults aggregating and reproducing in new areas by implementing new behaviors. Finding a new area where reproduction is sustainable may be a slow process of trial and error for the population. Often the colonization process undertaken by new recruits into depleted areas fails because of entrained homing behaviors.[47]

If the Aleutian Basin pollock was a self-sustaining population and if experience with closely related species holds, then successful recolonization and rebuilding of the depleted Donut Hole pollock population might be expected to take considerable time. Strong year classes in the Aleutian Basin appear to be rare, and depletion of the spawning population by overharvesting may be detrimental. In the case of the northern cod, many local spawning groups were decimated and they haven't rebounded some twenty years later. In the Gulf of Maine, the coastal cod collapsed in the 1940s, and at forty of ninety coastal spawning locations they have not reappeared despite the passage of over a half century.[48] In other words, the population has yet to recover.[49] Only recently is there sign that some cod stocks are starting to reappear.[50] Furthermore, the Bohuslän herring has been absent from coastal Sweden for over a century.[51] It appears that some populations, once decimated, do not quickly recover.

### The Empty Hole

The loss of almost 13 million tons of pollock from the Donut Hole over several years is equivalent to about 20 billion fish, or about three fish for every person on Earth. It is an unrecognized tragedy. But the social impact of the pollock collapse in the Aleutian Basin is hard to compare with the collapses of the sardine and northern cod for several reasons. Perhaps the Aleutian Basin pollock received little attention in spite of the dramatic population decline because the fishery was far out in the high seas where it had a relatively small impact on coastal communities. From socioeconomic and cultural points of view, pollock in the Donut Hole wasn't a centerpiece of American commerce, it was a foreign fishery. Pollock didn't provide subsistence to Native American fishermen or employ recently settled immigrant populations. Furthermore, any seabird and mammal populations in the Donut Hole feeding on pollock and impacted by the collapse of the fishery were largely invisible to human observation because of their remoteness, out of sight and out of mind.

In 2007, a test fishery conducted under the auspices of the Central Bering Treaty explored the central Bering Sea; a pair of Korean trawlers spent two weeks trawling for pollock in the Aleutian Basin. They caught just two pollock. The major species of fish in the catches was the homely "smooth lumpsucker."[52] After twenty years the population of pollock still has not recovered.

# **6** Viking Invasion: Norway's Link to the Pollock Industry

*The farm laborers lived in the little fishermen's cottages down by the steel-gray fjord, each with a small piece of land about it. They were pledged to work many weeks of the year on the big farm, and cultivated their own land when they came home in the evening, and even then they had to resort to the sea for their principal means of subsistence. They took part in the herring fisheries in the autumn, and in the winter sailed hundreds of miles in open boats up to Lofoten, perhaps tempted by the hope of gain, but perhaps, too, because on the sea they were free men.*

JOHAN BOJER, in *The Last of the Vikings*

I had heard the words, "You should talk with John Sjong" again and again as I spoke with people about the pollock fishery, so I was eager to interview him. He was a "Highliner" fisherman and a successful businessman. His story personifies that of many other successful Norwegian-American fishermen. I drove to his house in an affluent neighborhood of Seattle overlooking the Puget Sound, and turned onto his street. As I searched for his house, it became obvious which house was his—the one with the ship's mast grounded in the garden. I walked up to his front door, glanced at the bumper sticker advertising support for a certain political candidate on his Volvo, and thought "uh oh, we may not get along."

John opened the door, releasing the laughter of young children playing inside. "Doin' some babysittin' wit' the grandchildren," he offered in a thick rhythmic accent. I was put at ease. John was in his seventies, fit and alert, and still an avid skier. He had a medium build, a broad stern face, and salt-and-pepper hair. His steady eyes and thick black eyebrows struck me immediately. I had seen this confident gaze in the eyes of other skippers.

We sat down in his study. Everything was carefully placed. There were two mugs and a pot of strong coffee on the table between us. A platter of spritz cookies baked by his wife, Berit, nestled in the middle. Polished brass instruments from his last ship, *The Pavlof*, were displayed on the wall, and a hand-worn wooden steering wheel was mounted in the floor in front of the window. It could have been his wheelhouse, but this was his den. It felt like a bear's lair.[1]

Sjong is a model of the Seattle-based fishermen who immigrated from the central west coast of Norway. They often had minimal education, but made up for that shortcoming with hard work and native intelligence. Fishermen from that area are known to be industrious, hardworking, and opportunistic.[2] They are sometimes referred to as the "Scots of Norway." When fish populations were depleted off their own coast they would travel to distant waters to fish, often to Iceland, Greenland, or even farther out. Their arrival on the West Coast of North America was an expansion of that culture.

For the first few minutes of our talk, I thought John's expression revealed some discomfort and perhaps a little impatience. He shifted in his chair as if being interviewed was something he didn't enjoy. It wasn't natural to him, but he'd put up with it—as he might endure sitting on a hard wooden pew with his wife for a Sunday sermon at the Lutheran church. An interview is one of life's conditions that you endure with a brave face—like a visit to the dentist. But soon he relaxed and settled into the pace of the interview. He faced me solidly, but not quite squarely. There was something familiar about the way his body was cocked in the chair, and afterwards I realized it was a posture I'd seen before in a ship's captain as he observes his charge from the bridge; the skipper's chair is often placed to one side of the cabin, so that his body turns slightly toward center. Then while one eye peers at the sea in front, the other is able to steal a glance behind at the work on deck. He is alert and watchful.

In our discussions, Sjong acknowledged the risks he faced at sea and in the fishing business, but he said, "I had solid boats and good people, and never really thought of it as dangerous." When he was fishing crab in the Bering Sea, in the morning he would leave a calm anchorage in the Aleutian Islands to head into the fog, not able to see what lay ahead of him. I thought it must have been this way for the Vikings, preparing carefully for a voyage, setting off for months, not knowing what the sea had in store for them, and not foretelling the fate their families would endure if they didn't return. Even now in Norway, the older sections of the cemeteries in the cod fishing region of the Lofoten Islands are dominated by the graves of women and children; the men were usually lost at sea instead.[3]

Fishing is the most dangerous occupation in the United States[4] (see Sidebar 2). The Bering Sea is one of the most treacherous bodies of water in the world, a center of global storm activity. Gale-force winds blow in along the Aleutian storm track and tend to linger awhile near the productive fishing grounds. "The price of not learning the ways of the sea is death."[5] Yet the Norwegians weren't too concerned; they loved the life at sea.

Respect, but not fear, of the sea is a sense I got in talking with Sjong. You realize you are taking risks but you do what you have to do. Kaare Ness, a fisherman from Norway who is the cofounder of Trident Seafoods, said it like this, "I didn't consider fishing dangerous, but of course it is. I just thought it wasn't. [I] was never scared of it. When I think of all the men who died, I know now how dangerous it is."[6]

Sjong had confidence in his own ability and in his boat. I felt this about the course of the interview as well. In the setting of his den, he had control over the direction we would take. When our interview was over and I was leaving, a young teenage girl, his oldest granddaughter, appeared at the door. He spoke in his Norwegian sing-song, "Megan, do you know it's been a month since I've seen you?" as if *her* absence made him suffer. John had retired from his life at sea and it was a poignant statement. Now he's the one who worries about his absent loved ones.

Fishermen from Norway like John Sjong and other countries (see Sidebar 3) played an important role in developing the groundfish industry of Alaska. How did that come to be? What were the circumstances underlying their success?

Sjong's personal history reflects many attributes of the Norwegian immigrant fishermen. As a youth he lived in the village of Sykkylven, near Alesund on the northwest coast of Norway, and left school at the end of eighth grade. Some years later, he was apprenticed to a machinist and planned to be an engineer, when a friend invited him to immigrate to the United States. The friend got engaged and never made it; John left his home anyway and arrived in the United States in 1960. When asked why he left Norway, Sjong says, "Well, you have to understand that in those days there wasn't a lot of opportunity there. That was before the [North Sea] oil.'"[7] Another old Norwegian fisherman told me of his early life in Norway after World War II, "Norway is nothing but rocks—we were struggling to survive. There was nothing to feed us. I was fourteen years old when my mother put a hat on my head, opened the door, and told me to go out and make my way because they couldn't feed me any longer." These expressions of the hardness of life in those times stick to you like herring scales. Life has changed a lot in Norway after the gush of cash from the oil fields started flowing in the 1970s. Now Norway is one of the world's richest countries, and people are trying to get in rather than get out.

Sjong was ready to live the American Dream, but not quite yet. As part of his immigration process, he spent two years in the US Army where he learned English. He said that he spoke very nice English afterwards, but when he got back to Seattle's Scandinavian community of Ballard and the wharfs, where Norwegian was almost the native tongue, "It went to hell again."[8]

Sjong hired on as a mechanic on a crab boat fishing out of Kodiak. In those days, fishing for red king crab was lucrative. He saved his earnings, and he came to own several of his own boats. When the collapse of the king crab stocks happened in the early 1980s, Sjong and some of the other Highliners were ready for it. They had already converted to trawling or had vessels that could either fish for crab or trawl.

The FCMA of 1976 opened a huge opportunity as Americans had first priority to a huge biomass of fish in the Bering Sea. It was first come, first served to whoever had the boats to catch them. But US banks were not interested in financing the $10 million–$30 million for each boat needed to catch and process the fish.[9] They saw the risk but didn't recognize the reward. Instead, many fishermen turned to Norway, which had vast experience in the business of fisheries.

### The Norwegian Brotherhood

Part of the success of Norwegian immigrants in the fisheries of the Bering Sea is due to the Norwegian brotherhood. Norwegians tend to pull together and help each other.[10] In his book *The Last of the Vikings*, Johan Bojer describes the great risks, dangers, competition, and hardships of fishing, as well as the camaraderie and the tradition of "fronting." That is, loans were made from person to person to buy boats and gear on the basis of hope for a good season. In the early days of the pollock fishery, a lot of business was done on a handshake.[11]

As in the old country, fishermen sometimes risked their own earnings on the ability of others to also catch fish. Norwegian businesses often extended large amounts of credit to reputable fishermen.[12] On the other hand, there were complaints from other Americans that the Norwegians would mostly hire only other Norwegians as crew.

There was a critical connection between the immigrants and financial resources in Norway.[13] Norwegian investors financed a massive rebuilding of US ships into trawlers, factory trawlers, and processors, with much of the work done in Norway and financed by Norwegian interests.[14] Norway was flush with cash from the rapidly developing North Sea oil industry, half of which was owned by the state company Statoil. The ship building industry of Norway was spread across fjords in rural areas along the coast. The industry was large and struggling. In the 1970s and 80s there was intense competition in the shipbuilding industry with Asia. Many jobs were lost. The government used funding from their oil profits to subsidize and shore up the ship-building industry, partly to stop the flow of workers into urban centers.

One mechanism to promote ship building was to give low-interest loans and grants that covered the equity. There was a huge amount of money at stake as the typical cost of converting a ship was $38 million.[15] Norway's Christiana Bank alone had a loan portfolio in the Seattle-based fishery of $435 million, and other Norwegian banks, such as Creditbank and Bergen Bank, also had large outstanding loans.[16] The total investment of Norway's banks was probably around $1 billion. Norway wasn't the only investor. There were also US government subsidies and Korean and Japanese investments to rebuild and modernize the fleet. Often a "package deal" was put together, with the loan, architect, and shipyard bundled together.[17]

John Sjong and his partner, Konrad Uri, started the groundfish rush in the Bering Sea.[18] They bought the *Sea Freeze Atlantic* in 1979 for $2 million, in part with Norwegian equity, and spent about $4 million retrofitting the ship. The *Sea Freeze Atlantic* was the first factory trawler originally built in the United States. She was constructed in Baltimore in 1968, funded by US government subsidies, to explore the fishing grounds off the coast of New England, Canada, Greenland, and

Norway. The Bureau of Commercial Fisheries and the Maritime Administration released funds in 1968–69 to pay half of the $5.3 million cost.

The original *Sea Freeze Atlantic* was 295 feet in length and carried a crew of ninety. But there were technology problems and trouble finding a crew willing to spend months at sea. The ship was laid up after two years, and sold at a loss in 1974. Then in 1979, two fishermen in Norway looking for opportunities abroad came along. Father and son, Michael and Erik Breivik, provided substantial equity and they partnered with Uri and Sjong's company, Trans Pacific Industrial, to buy the ship. The ship was partly retrofitted in Norfolk in 1980, and more work was done in Seattle. She was renamed the *Arctic Trawler*, and a crew was hired, including Kjell Inge Røkke and eight other members from Norway. Breivik couldn't officially captain the ship because the Jones Act required an American citizen with a captain's license, so he became the "fishing skipper" and young Røkke was the bosun.

For the first couple months in the Bering Sea, they didn't catch much. Sjong said, "We were chewin' nails and wonderin' what the hell was going on."[19] Then they ran into fish, lots of fish. But they had problems marketing them because the buyers were used to dealing with Atlantic cod, even though the Pacific cod was an attractive product for the whitefish market. In those days, "frozen fish was a bad word."[20] In 1981 the *Arctic Trawler* reportedly landed 4.65 million pounds of true cod fillets in Seattle.[21] Eventually they fitted the *Arctic Trawler* with fillet machines suitable for pollock and they kept the ship for seven years. With regard to his pioneering fisheries ventures being lucrative businesses, Sjong admitted, "We made a little money, but it isn't always the best to be number one. A close second is much cheaper."[22] The *Arctic Trawler* was later sold to Japanese investors forming a company called Arctic King Fisheries.[23]

Sjong, Uri, and other partners bought the *Arctic Trawler*'s sister ship, the *Sea Freeze Pacific*, which had earlier been converted into a crab processor. This was the first vessel brought to Norway for retrofitting into a pollock factory trawler in 1985. Sjong and partners renamed her the *Royal Sea* and formed the Royal Seafoods Company.[24] They converted four more vessels to pollock factory trawlers in Norwegian shipyards. Marketing surveys showed that restaurant customers readily accepted Alaska pollock as a substitute for cod, leading to one of the big steps in the development of the American pollock industry. In 1986 Sjong announced that the Skipper's fast food chain had agreed to buy pollock as a replacement for cod in its fish-and-chips product.[25]

Sjong sold Royal Seafoods and formed a new company called Regal Seafoods. The new owners of Royal Seafoods went bankrupt and the company was taken over by a California debt management firm called Trust Company of the West. Later Royal Seafoods was taken over by Resources Group International (RGI).[26] Regal Seafoods ran six trawlers. Eventually Sjong tired of running a big company and sold Regal to New York investors. This venture didn't fare well for the new owners, who sold out. Several of their boats were bought up by American Seafoods.

Erik Breivik left the *Arctic Trawler* to found Glacier Fish Company in 1982, of which he is still CEO. Breivik was born in the Lofoten Islands and moved south to Alesund with his family when he was thirteen years old. His father started jigging for cod in a sixteen-foot open boat in 1932 and by 1973 had built a 245-foot factory trawler, the largest in Norway at the time. Breivik started fishing for his father in Norway when he was fifteen and within a few years he was a skipper.[27] His voyages took him to the Barents Sea and to the waters around Greenland and Iceland. It wasn't uncommon to be gone from home for four months on fishing trips.

In the late 1970s he became increasingly disillusioned with allocation disputes and quota limitations in Norway and continued his westward migration across the oceans to the Bering Sea, where American fisheries were developing. It seemed like a natural extension and there were opportunities for growth there. In 1978 he was contacted by John Sjong, and they met with the marine architect Lars Eldøy to plan the *Arctic Trawler* conversion, which was completed in 1980.

After he formed Glacier Fish Company, in 1983 he built the *Northern Glacier*, a state-of-the-art vessel made especially for catching and processing groundfish in Alaskan waters. The ship was built in Tacoma but partly financed by a Norwegian bank. Then in 1988 he had a mudboat called the *Magnus Sea* rebuilt in Norway, with design assistance from Eldøy, to become the other main pollock fishing vessel in his fleet, under a new name, *Pacific Glacier*.

Breivik is a solid man with pale blue eyes and powerful convictions, which he seems to tone down to be polite. He stresses the advantage of having a strong mind and of not giving up in the face of adversity. Breivik says, "I guess hard work never hurt anyone" and "Don't give up and something good will happen." It is important for a fisherman, he says, to have the ability to "smell where the fish are."[28] One could interpret this ability as a fisherman's intuition, or the abstract process of putting together his experience and knowledge to give rise to a sense of where to deploy his nets. Breivik has a reputation for running a first-class operation in the fleet, with attention to detail and safety, and producing a top product. It is remarkable that out of the eight Norwegian crew that signed on to fish with him on the *Arctic Trawler* in 1980, three are still working for him at Glacier Fish over thirty years later.

John Sjong and Erik Breivik had pivotal roles as the pioneers of the US pollock industry. Besides their role in developing the fishery, they provided opportunities for others to enter the fishery, including Kjell Inge Røkke.

## Røkke and American Seafoods

Kjell Inge Røkke was the original bosun on the *Arctic Trawler* and his story is a remarkable example of rags to riches. Røkke grew up in Molde on the northwest coast of Norway, was academically disadvantaged with dyslexia, and dropped out of high school at the age of sixteen. His parents were disillusioned and thought he was headed no place. His teachers gave up on him. He was drifting. One morning his mother woke him up and said, "You are going to work." She had secured him a job on a fishing boat. It suited him. By the time he was seventeen years old, he was fishing for cod in the Barents Sea on a trawler named the *F/T Svalbard*, based out of Svalbard, Norway.[29] He started at the very bottom rung on the ship as an engine room oiler but showed he was a hard worker with ambition and worked his way up to the deck.[30] Eventually he met Breivik in the English port of Grimsby and was offered a job on one of his boats. A few years later he was recruited by Breivik to fish for pollock in the eastern Bering Sea.

Røkke arrived in America with big plans for his future, a dream of owning his own boat, being his own master, and making a lot of money.[31] He arrived with "hardly two pennies to rub together." He saved his earnings, and in 1982 he bought his first trawler, the *Karina*.

The *Karina* was a wide-beam seventy-foot vessel with a low center of gravity. She was modeled after a Norwegian design and was reported to be able to right herself after a seventy-five degree roll. Soon he added the *Karina Explorer*, which sank in 1986 with the loss of all hands. But the challenge of catching enough fish and marketing them was a rocky start, and he had a lot of unhappy creditors breathing down his neck.

Later he formed a business partnership with Bob Breskovich, who was also his mentor, and they bought the crabbers *Centurion* and *Vanguard*. They took advantage of a depressed market for crabbing boats. They rebuilt the ships as trawlers and renamed them the *Aleutian Harvester* and *Aleutian Challenger*, at 94 and 86 feet, respectively. Later they added the 120-foot *Ocean Leader* to their fleet.

The *Regulus* was built in 1972 in West Germany as a 100-meter supertrawler. In 1982 it was bought and rebuilt as a fish processor by Alaska Brands Corporation and Inlaks Seafood, and renamed the *Golden Alaska*. Inlaks was a group of Anglo-Indian investors. The *Regulus* couldn't trawl in US waters because of the Jones Act requirements for American construction, ownership, and crews, so it served as a mothership to three catcher boats. This venture wasn't successful, so for a while she was tied up and rusting in Seattle. There was a rumor she was headed to Nigeria. In 1984, Assen Nicolov sold the *Golden Alaska* to Kjell Inge Røkke and Bob Breskovich for $3.5 million, and they retrofitted it to process pollock. Reflagging the *Regulus* as an American ship was noted as a huge success in the process of Americanization of the fishery. The rebirth of the *Golden Alaska* was considered a step forward against the Japanese hegemony of the pollock fishery. Senator Warren Magnuson celebrated the launch, pointing to the ship, "This is what I have worked for all my legislative life, to allow the large factory trawlers to come in and employ American people, fly the American flag, and catch the fish with American deckhands."[32] Of course, Røkke was actually Norwegian. He never did gain US citizenship.

Around this time there was a tragic and disturbing incident that haunts Røkke's past. On November 24, 1985, Røkke was on the *Golden Alaska* from which he ruled his other fishing boats. The crew said that nothing happened without Røkke's knowledge and approval.[33] The ships were fishing just south of Unalaska Island. The usual captain of the *Aleutian Harvester* missed the trip and a younger and less experienced fisherman was filling in for him. A big storm blew in and the skipper of the *Aleutian Harvester* radioed in to Røkke, saying that he wanted to stop fishing and look for shelter. By this time the swells were about thirty feet high.

Gunnar Stavrum reported in his book that Røkke didn't answer the captain of the *Aleutian Harvester*, which seemed to signal from Røkke a lack of approval.[34] Because of the debts Røkke had taken on there was a lot of pressure to keep the fish deliveries coming to the processor. The inexperienced skipper gave in to the pressure. At 7:30 a.m. the captain of the *Golden Alaska* woke Røkke and told him that the *Aleutian Harvester* had disappeared from radar and they had lost radio contact with her. Røkke seemed nonchalant and did not report the missing ship to the Coast Guard until ten hours after the *Harvester* had disappeared. The *Golden Alaska* was reported not to have conducted a search for the missing ship. Instead the *Arctic Trawler*, which was nearby, did a search but without success at finding the ship or survivors. The three crew members on the smaller vessel perished; one body later washed up in the Aleutian Islands.

During a later Coast Guard inquiry, Røkke said that the captain of the *Aleutian Harvester* was responsible for his own decisions, while the actions of the *Golden Alaska* were the responsibility of its own captain. Others indicated that shoving aside responsibility for decisions wasn't like Røkke. Furthermore, there were allegations that the *Aleutian Harvester* was not seaworthy for conditions in the open ocean of the North Pacific. It had a low square stern with twin screws, a configuration known to be treacherous in a heavy following sea. However, Røkke was not found negligent by the inquiry, but was instead fined $5,000 for not following appropriate procedures. To be fair, Røkke has fierce supporters over this incident. They say Røkke "would not ask someone to do something that he himself wouldn't do." They also report that Røkke was generous and concerned about his employees. However, this unfortunate incident reflects the focus of Røkke to catch fish, take risks, and make money above other considerations in these early years.

By 1986, business had returned to normal and the *Golden Alaska* was making a lot of money. The *Aleutian Harvester* was replaced by the 123-foot *Ocean Beauty*. Breskovich and Røkke made a good team, with the former providing strategic planning and contacts, and the latter contributing explosive energy. Breskovich remarked that every year there are thousands of Norwegians who come to Seattle, but there is only one Kjell Inge Røkke.[35] Eventually their different approaches conflicted, and a year later Røkke split with Breskovich and sold his share in the *Golden Alaska* to Nichiro Gyogo Kaisha Ltd. along with his share in the *American Beauty* and the *Ocean Leader*. He sold his share in the *Aleutian Challenger* to Breskovich. That year he founded American Seafoods and in quick succession used Norwegian shipyards and financing to convert three hulls into the factory trawlers. He bought the US supply ship *Maureen Sea* in Singapore, took it to Norway, and converted it to the *American Empress* for $38 million. The US supply vessel *Artablaze* was converted to the *American Dynasty* for $42.5 million. Finally, the 74-foot research vessel *Acona* was converted to the *American Triumph*.[36] The plan for converting the *Acona* was actually carried out by PanPac, a New Zealand company owned by the Fletcher Challenge Group. They rebuilt the boat in cooperation with American Seafoods at a cost of $40 million. In 1990, just after the boat was delivered, it was bought outright by American Seafoods.

Røkke was one of the earliest pollock fishermen to take American-built hulls from old tankers and Gulf of Mexico mudboats and convert them to pollock factory trawlers and processors. Røkke said one of the reasons he contracted the vessels and formed ASC was "to take advantage of the US legislation to allow for US vessels to be converted abroad to obtain more attractive pricing and financing and a better quality vessel."[37]

The venture gambled on the success of the Alaska pollock fishery, and it was a gamble that paid off. Røkke's remarkable story had launched. But Breskovich's luck had run out. In 1989 the *Golden Alaska* caught fire while tied up in downtown Seattle. The city deployed 125 firefighters to fight the blaze. The ship began listing to port ten degrees when water from all the firehoses accumulated on the top deck, and the ship was in danger of capsizing. Sixteen hours later, the fire was tamed. The *Golden Alaska* was a total loss and was not adequately insured.[38]

Røkke's three new boats were state-of-the-art trawlers and equipped to process surimi. He had made a good decision to process surimi, as the price was sky high, and his top-notch boats were able to outfish his competitors in the race to catch fish. He got rich in his first season.

The stream of money was flowing. Røkke was on a roll. He began to acquire the assets of other companies as they faltered in the tricky market for groundfish; eventually Røkke's company controlled 40% of the American pollock harvest. Røkke also controlled 45% of the pollock quota in Russia and owned part of a Russian company fishing for pollock. Greenpeace and other sources reported that 40% of the biomass of Russian pollock was being harvested each year, and it was an example of serious overfishing.

Later Røkke was forced to sell American Seafoods because of anti-foreign ownership legislation initiated by Senator Ted Stevens. He went on to found Aker-RGI, a holding company that controlled 10% of the world's whitefish supply and was one of the world's largest fishing companies.

Since his American fishing days, Røkke has become one of the world's wealthiest men, attaining a fortune of $3 billion, and along the way attaining the reputation as a "bad boy" of marine fisheries conservation. His company American Seafoods was involved in enterprises from Russia to Argentina that were accused of overfishing, and in the past the company was pilloried by Greenpeace as a group of "ocean nomads."[39] Whereas the classic fisherman fished where he lived and passed his knowledge and the resource on to the next generation, the new factory trawlers, epitomized by Røkke's fleet, just kept moving to more distant oceans to fish. He was characterized as a ruthless predator. Speaking for himself, Røkke at that time considered that it was not his job to protect the resource; it was his job to follow the rules that agencies had established.[40] The problem was that in some seas there were no rules to follow.

Kjell Inge Røkke is an enigmatic and controversial figure in Norway. To some he is a flamboyant tycoon and anti-hero. He drives speedboats and fast cars, ignores rules, and flaunts his wealth and influence. He's been called a modern oligarch.[41] In other eyes he is a modern day folk hero, a lucky hardworking guy who found opportunity in the New World and became a successful entrepreneur. Like John Sjong said, "The harder you work, the better your luck gets." The conflicting public opinions of Røkke point out there is a fine line between the roles of protagonist and antagonist.

Røkke has been described as a man with a lot of will, energy, and ideas.[42] "He is a man willing to take huge risks."[43] He is said to have an iron will and "unorthodox" methods. Those are traits that make a good fisherman, and a good businessman. His success drew the attention of several powerful enemies, namely Greenpeace and Senator Ted Stevens, leading to his downfall in the United States but facilitating the launch of his international career. His empire has branched out into offshore oil drilling and his companies have owned many other ventures, including clothing producer Helly Hansen, athletic shoemaker Brooks Sports, and Aker Philadelphia Shipyard that builds oil tankers. The shipbuilding arm of his empire is constructing the world's largest passenger ship, *Project Genesis,* at a monstrous length of 360 meters, nearly a quarter mile long.

In 2005, he was prosecuted on a charge of corruption in Norway for bribing an official to falsify a captain's license, and served but 25 days of a 120-day sentence before being paroled.[44] Ironically, the next year he was chosen by the Norwegian Parliament to receive the prestigious Peer Gynt Award, given in recognition of making positive contributions to society. Previous winners of the award were distinguished human rights activists, diplomats, musicians, and artists. Afterwards, a faction of activists demanded that the award be retracted because one of his companies was allegedly providing support services to US Guantanamo Naval Base.[45]

In 2008, one of his companies, Aker-Biomarine, announced a partnership with the World Wildlife Fund and gave them one million Norwegian kroner. Soon Røkke was being praised for his conservation efforts, a remarkable turnaround. Røkke's Aker Seafoods company says it is now leading the charge against illegal fishing, in part because it affects their own quota and drives market prices down.[46] What a paradox that one of the big fishing companies is also a major force for conservation!

Another Norwegian, Bernt Bodal, took over from Røkke as president and later chairman and CEO of American Seafoods. Bodal also started off as an immigrant fisherman in the Bering Sea but, unlike Røkke, had become a US citizen. Bodal has a colorful past. After Bodal graduated from high school in Norway in the mid-1970s, he toured Europe for three years in a progressive rock band named Høst. The band enjoys near-legendary status in Norway and has done reunion concert tours. Bodal has recently played gigs and held his own with Eddie Vedder of Pearl Jam and Roger Daltry of The Who. Away from his day job, he is a current member of the White Sox Allstars, a reunion group of rock legends that includes members of Journey and Queen. Bodal himself has the look of a clean-cut rock star, maybe a taller Nordic version of Eric Clapton. In his office an electric guitar hangs on the wall.[47]

Bodal is different from the other immigrant fisherman because he isn't from the west coast, but from near Oslo. He grew up hearing stories about his uncles who had emigrated and was fascinated by life in America. In 1978 he borrowed a thousand dollars for the plane ticket and came to the United States without an extra penny. He got a job on a crab boat and later became a partner in a crabber/trawler in a joint venture fishery. In 1989 Bodal started working with Røkke at American Seafoods.

It's ironic that Bodal's uncles told his mother that Bernt was a nice guy and a good worker, but "he is never going to be a fisherman and he should look for another profession." But as Bodal eventually told a reporter in a strong Norwegian accent, "I've always been competitive in everything I do."[48] He ended up skippering fishing boats in the Bering Sea for thirteen years and became the principal owner and CEO of one of the largest fishing companies in the United States. He says that the management skills he needed to run a major company were learned, not in school, but as a captain on a fishing ship where he was responsible for running the ship, maintaining the welfare of the crew, and making the catch. A captain often has to make life or death decisions instantly. He says what makes a good fisherman are "risk, hard work, and quick decisions."

Bodal told a story about being caught in a bad storm in Shelikof Strait during his days as a fishing boat skipper. It was 10 degrees Fahrenheit and winds were gusting to nearly 100 knots. The seawater spray was freezing on everything. As he tells the story, the faraway look in his blue eyes leave no doubt that he is reliving the memory. Freezing spray is a dangerous situation on a crab boat as the pots are metal and there is a huge surface area for ice to build up. This is the circumstance when the boats become top heavy and unstable in the swell. Two other crab boats flipped over that night. Bodal and the crew were concerned and tense, but his fishing experience had taught him to never panic, to always stay cool and think things through. That night it took them six or seven hours of slow running through the wind and waves to get into a sheltered bay. They ended up chipping ice off the anchor and chain for another four or five hours so they could deploy it.

Common themes weave through the stories of the Norwegian fishermen. I've presented only a few, but there are many others, of young undereducated Norwegian immigrants escaping a limited future in the Old World and finding great success in the pollock fishing industry. Some of the old fishermen retired wealthy, others lost their fortunes and didn't end their careers in prosperity. In talking with them, I sometimes got the sense that they were lost on the shore. They were more at home on their boats. They had tans laid down by the sun, then permanently pressed into skin by the wind. The eyes of many were pale blue as if the sea had sucked the color from them. They spoke slowly and deliberately in a deep friendly sing-song.

Most of the fishermen came after the war from the areas of Alesund and Molde on the northwest coast of Norway or the small island of Karmøy, which is farther south between Stavanger and Bergen. The island was known for the herring industry and as the seat of the Viking chieftains.[49] I heard of so many fishermen and families from Karmøy, it made me wonder if there is anyone left behind. Kaare Ness, fisherman and cofounder of Trident Seafoods, said, "I was a Karmøy man, and Karmøy men fished because that was what they knew. . . . They had no choices. Everyone was very poor. As soon as we were fourteen or fifteen, we had to work."[50] Now the Karmøy club even has a local chapter in Seattle's neighborhood of Ballard for the island's emigrants. Many of the old Norwegian-American fishermen, and even some non-Norwegian fishermen (called "Norwegian wannabes"), have summer homes in Karmøy now.

What made these young immigrants from Norway successful? What led them to take the risks that they did, both at sea and in the business world? Was it family and cultural history? Did they find freedom from the limitations of a tightly knotted society? Or by virtue of having left Norway for an uncertain fate, were they risk takers by nature? Once they had taken the chance to emigrate, did their momentum carry them forward? Or did they figure they had nothing to lose?

The most successful fishermen were also very intelligent businessmen who could see when the amount of competition for a resource was too great to sustain itself under the pressure, and they moved to another resource. John Sjong moved from crab to cod to pollock, always on the move for new ideas. At the time he shifted to fishing pollock, "they seemed unlimited." Sjong said, "Competition was the driving force. I wanted to be the top boat."

Kaare Ness also remarked that ambition and competition drove the top fishermen. He always wanted to catch more fish than the other guy. He never thought about the money, though; it was all about how many fish you were catching that was driving you.[51] Ness was more than an able competitor. With Chuck Bundrant and Mike Jacobson he founded Trident Seafoods in 1973, and it has grown to be the largest fishing company in the United States with forty vessels and 4,000 employees. Ness's wealth is estimated in the hundreds of millions.[52]

Pollock represented a great fishing opportunity. The opinion that pollock were unlimited was not far behind what scientists were thinking at the time. Sjong finally left the pollock industry when he saw that the fishery had become overcapitalized and profitability was lagging. The company that bought him out went bankrupt. Many of their ships were opportunistically gobbled up by Røkke's American Seafoods. Røkke was an innovator and driven to success. Røkke is said to have seen the opportunity to buy distressed ships as a unique chance because he foresaw that the fishery would be rationalized, and quotas would be granted based on vessel catch history. He guessed right.

In 1944 Graham wrote in *The Fish Gate*, "Fishing skippers may be famous before they are thirty. Surely there is no other calling that gives such good financial rewards to men of so little schooling; but schooling hardly counts—-the rewards are only for the able, the persistent, and the men who can command the respect of a lively crew."

Tor Tollessen is a respected personality in the Seattle fisheries community. An immigrant from Karmøy, he fished for nineteen seasons, opened a successful marine electronics business, and is well connected with almost everyone in the business. Tollessen says there are three characteristics that make a good fisherman: drive, intelligence, and luck. If you have all three of those, you are a successful Highliner. If you have two of the three, you make a good living. If you have only one of these characters, you will end up selling your boat. And what about risk? "You create luck by taking a risk."[53]

Another icon of West Coast fisheries science, Lee Alverson, said

"that the most successful fishermen, the Highliners, were independent thinkers, willing to take risks. And they had a great love for going
to sea."[54]

At the time these immigrants made their move, there was an
alignment of the stars that gave them opportunity. Their independence, adventurous spirit, and willingness to take risks as individual
fishermen-innovators played a large role in their success. The FCMA
created a great opportunity. The establishment of the 200-mile limit
and ejection of the foreign fisheries created a vacuum that sucked
those nearby into a position to claim the riches for themselves. A pile
of fish was there in the Bering Sea available to be scooped up on a first
come, first served basis. The resource was huge. The immigrants had
strong connections in Norway and the Norwegian government was
flush with cash from the North Sea oil profits. The Oslo government
supported programs to shore up the numerous shipyards located in
the fjords of Norway with guaranteed and low-interest loans. All of
these factors came into play opportunely for a few lucky individuals.
The fearless Vikings were among those at the front of the charge to
"Americanize" the pollock fishery.

What was it about these men that made them take the risk of fishing, a dangerous profession in a dangerous sea? There is something
in the nature of Norwegians, the lure of tradition rooted in roman-
tic tales of the Vikings. The Vikings roamed the world with a sense of
freedom and a reputation for being fearless and strong individuals.
Later, their ancestors fished the huge stocks of cod around the Lofoten Islands and fished herring closer to home with the same sense of
adventure, freedom, and pursuit of opportunity. This was a tradition
that passed from father to son. The same cultural tradition and sense
of family heritage can be found in fishermen today.

· · · · ·

SIDEBAR 2: TRAGEDY AT SEA

*Things can go from bad to worse very quickly on a ship at sea, and a tragedy
can unfold in the blink of an eye. This is particularly the case in the Bering
Sea, where wind and storms—and dangerous icing conditions—are common. It is no wonder that working on a fishing boat in the Bering is called
one of the world's most dangerous jobs. In an average year, say 1992, it was
reported that thirty-five fishermen died at sea and forty-four boats made*

one-way trips in the Bering.[55] But even in calm water, the Bering Sea can be treacherous.

The Aleutian Enterprise fished for ten years in the Bering Sea, experiencing terrific storms and dangerous conditions, before sinking in calm weather on March 22, 1990. Nine crew were lost in the mishap, and twenty-two lucky ones were plucked from the water alive to survive the tragedy.

At the time of the sinking, Arctic Alaska Fisheries Corp. was the nation's largest fishing company. The Aleutian Enterprise was built in 1980 in Massachusetts with the intent of being a crabber, but litigation kept her tied up for three years. In 1983 she was bought by Aleutian Enterprise Ltd., said to be owned by a group of California investors, and was converted to a groundfish catcher-processor in a shipyard in Alabama. The ship was 143 feet long and weighed 133 net tons. Arctic Alaska Fisheries was operating the ship to fish for cod and pollock in the Bering Sea.

On the day of the accident, the ship was listing slightly to port and had a hold full of fish, as a trawl filled with seventy tons of fish was coming onboard. The trawl was overfull, and as it came on deck, the fish were compressed into the rear of it splitting the net open. The slippery fish spilled out of the rip and flowed downslope to the port side. The ship began to list even more. Several of the crew thought the ship was turning hard. Instead a chain reaction was leading to disaster.

Bob McCord was on his first trip as a fisheries observer. He was thirty-five years old, from Colorado, and trained in forestry. At the time things started to happen, Bob was working on the aft deck, taking measurements and monitoring the species composition of the catch. From his position he could see a red light flashing, which meant that the bridge was contacting the engineering department.

In fact the bridge was calling the engineering department to see what they could do about the listing, and they had started to pump fuel to the starboard tanks, unaware of the real danger. There was an open hole in the hull at the level of the fish processing plant where fish offal and skin were slid into the sea. The hole was covered by a cookie sheet of metal rather than having a water tight hatch.

Four years earlier a regulation had been enacted regarding the placement of openings in the hull with regard to a "load line" that should be marked on the ship. The load line is a mark on the hull representing maximum surface water level to which a ship can be loaded. Any opening near the load line needed to have a water-tight door. But the owners of the ship didn't comply with the statute and the Coast Guard didn't enforce it. The previous year, the Coast Guard had a dispute about the load-line with another Arctic Alaska Fisheries ship. Washington State Congressman John Miller intervened on behalf of his constituents, and the Coast Guard dropped the matter. In fact, they had become pretty occupied with the Exxon Valdez oil spill of 1989 and were happy to let one more issue drop away.

As the ship was listing hard to port, the ocean swells were lifting open the flimsy cover in the fish plant, allowing water to gush in "like a fire hydrant," which was a positive feedback system accentuating the problem the ship was in. The ship's alarm system wasn't working, but at this point, most of the crew realized there was a bad thing happening and made a rush for the exits and survival suits. What ensued was a chaotic scene, with some people heroically trying to pass out suits and help others put them on, while some panicked individuals tried to grab suits from those that already had them.

The water temperature was 32 to 34 degrees Fahrenheit. Water is a pretty efficient diffuser of heat, and being in a total-body ice bath without insulation sucks all the heat from a body with a few minutes. Bob McCord was last seen in the water without a survival suit and perished. Another young victim was a nineteen-year-old who had been a crew member on another ship, decided that the life at sea wasn't for him, and was catching an unlucky ride on the Aleutian Enterprise to go back home.

From the time that the net came on board to the point where the ship capsized and sank, ten minutes had elapsed. Two boats fishing nearby were on the scene within twenty minutes and rescued twenty-two survivors. Families of the victims and the surviving crew filed claims for $100 million. Most were settled out of court. In 1994, fourteen officials of Arctic Alaska Fisheries, including the twenty-eight-year-old captain of the ship and the president of the company, were indicted on 100 counts of criminal negligence and manslaughter. Evidence was presented in court that the company covered up safety violations and had falsified records of qualifications of the crew. Thirteen defendants pleaded guilty, were fined, and were put on probation. The president and founding owner of Arctic Alaska Fisheries pleaded not guilty; the trial went to jury and he was acquitted.

The tragedy of the Aleutian Enterprise spurred hearings and changes in vessel safety. These were largely opposed by the fishing industry who feared the added expense and wanted to self regulate. It was also opposed by trial lawyers' associations. Tragedies at sea had been lucrative.

Passage of the American Fisheries Act and implementation of a catch share program, which ended the fishing derby and the race for fish, can also take credit for safer conditions at sea. The Bering Sea is still a dangerous place, but the fishermen don't feel compelled to go out fishing in bad weather to retrieve their fair share of the quota. The industry has developed and continues to support a fine personal safety training program through the North Pacific Vessel Owners Association. But many people still feel that vessel safety and implementation of Coast Guard inspections could be improved.

· · · · ·

### SIDEBAR 3  PORTUGUESE IN THE POLLOCK FISHERY

The Norwegians, although leading the charge, weren't the only ethnic group important in the pollock fishery. Among the others was a contingent of Portuguese fishermen.[56] The connections with the Norwegians and others are remarkable. By coincidence, Artur Dacruz Sr. had fished on the Sea Freeze Pacific, which Sjong and Breivik later acquired. He was fishing for a German captain who had been recruited to skipper the new ship in 1969, and was signed to a two-year contract. Eventually he ended up on the West Coast and he brought his family over. His son, Artur Jr., started fishing for sardines when he was ten years old. When the family moved to Seattle, he attended Ballard High School for a year. Then Artur Jr. went into commercial fishing and he hasn't looked back. In the early days of the joint venture he worked for Wally Pereyra's company, ProFish, catching pollock and delivering it to Korean processors. Now he co-owns a boat with Kaare Ness. Through the connections of family and friends, many Portuguese came over to join Dacruz in fishing for pollock. About two dozen Portuguese fishermen now fish in the pollock and cod trawl fleet. Almost all of them come from the small fishing village of Afurada. Artur Jr. says, "The Norwegians, the Portuguese, we are people who came from the sea. All we know is the sea. I was born not 100 meters from the ocean, and that's all I saw every day since I was a baby—boats and fishermen."[57] Artur continues, "After all these years, I still love fishing. I like the excitement, I like the friendship, I like the challenge. . . . Very few people understand exactly what we go through every day. We leave Dutch [Harbor] and we go north and have nothing but water before us."

· · · · ·

# 7 A New Fish on the Block: Advancing Knowledge of Pollock Biology

Leo Szilard said that he used to lie in his bath for hours and think about physics. He would have his best thoughts there. Szilard wasn't just a bathtub dreamer, he was the nuclear physicist responsible for conceiving the nuclear chain reaction. It led to his work on the atomic bomb in the Manhattan Project. He received a patent for nuclear reactors with Enrico Fermi. Szilard was brilliant. But in 1947, discouraged by the threat of catastrophic nuclear war, Szilard gave up the study of physics for biology. After he became a biologist, he complained that he could no longer enjoy a comfortable bath. When he thought about biology he found that he always had to get out of his bath to look up another fact.[1]

What does this story have to do with pollock population dynamics? As Szilard discovered, biology is all about complexity and details. Biology involves the interactions of many factors, and hierarchical levels at which they operate and feed back upon each other. The solutions to problems are not elegant. Biology is messy.

Making predictions of population abundance is one of the main objectives of fisheries biologists. Fisheries managers would like to know what is coming into the population next year when they set the harvest levels this year. Not only do predictions of future abundance trends help managers, but also they test our understanding of the system. Marine fish populations naturally fluctuate with changes in ocean conditions that influence their survival. Population dynamics of marine fishes are driven by the shifting balance of birth and death rates, as influenced by climate conditions, density-dependent effects, diseases, fishing pressure, and the abundance of predators. Fish populations respond especially to variations in the "birth rate," or the number of fish entering the population each year. The number of fish entering the harvestable portion of the population each year is termed *recruitment*.

Scientists who study fish populations like to use metaphors to visualize them. Over the years, I have heard the factors influencing the recruitment process compared with such mechanisms as valves controlling water flow in pipes and a series of on-off switches in an electrical circuit. A scientist named Alec MaCall created another simple metaphor for fish populations. Think of the number of fish in a population simplistically, like the level of liquid in a basin.[2] The depth and width of the basin, producing the volume, is like the carrying capacity. The level increases and decreases in response to processes influencing it.

We can borrow the basic concept of a basin from MacCall's metaphor and enhance it by adding waves. Imagine the waves are environmental processes that cause the level of fish to rise and fall through time. These processes are numerous, and vary over different time and space scales with different periods and amplitudes. They overlap with each other. Their history of passing resonates in the ecosystem. Some are strong and infrequent processes, like climate shifts and community shifts. Some processes may even occur over geologic time, like periods of glaciation ice or changes in landscape. On top of these long-term changes are processes with shorter frequencies, like decadal-scale regime shifts and El Niño events. Then there are seasonal cycles in climate, hurricanes, and storms. And those are just the physical events. One could imagine Leo Szilard in his bath thinking about this. Let's assume he is wearing a swimsuit. He moves his arm on one side of the basin slowly up and down, creating waves of low frequency but high amplitude. With only one wave generator, the pattern of water rising and falling is predictable, but varies in amplitude across space. He does the same with his arm on the other side of the basin. The waves collide and make a pattern on the surface. With a foot at the other end he creates a higher-frequency lower-amplitude wave, and with a big toe of the other foot he creates yet another even higher-frequency disturbance.

As you add wave generators, maybe just three of them, then the pattern from the surface looks chaotic. At any one point in the basin, the level of the water rising and falling can be measured by Szilard with a vertical ruler. It is unpredictable over time, and also differs in other areas of the tub depending on proximity to the source of disturbance. In reality the pattern may be very complex. If one measures the waves at a high frequency, the actual levels of rising and falling can be recorded. But if Szilard averages his measurements over space and time the record of amplitude dampens. If he measures at an interval that is greater than the period of the cycle, he may miss an accurate picture of the dynamics altogether. The scale of measurement is critical.

Viewing the basin from above the water rather than from the perspective at one point at the surface of the tub, he sees quite another pattern. Szilard observes that over a short period in one location, the water level may be influenced by a big slow wave passing, causing a slow rise and fall of the water level, but within that wave, smaller ripples can cause shorter-term increases in the water level. This is like a long-term trend in climate change being temporarily obscured by a short-term cooling event.

Fish populations respond to these environmental cycles. But it is more complicated than this simplistic imagery. In the real world, the light energy of the sun is converted into plant biomass under the influence of heat, turbulence, and nutrients. Plants are eaten by copepods and the energy is transferred up through the food chain, itself a complex process. Copepods reproduce and their young become prey of pollock larvae. Then there are other biological factors to consider that also influence population dynamics of pollock, like competitors, predators, and diseases. These may influence different life stages of the population being considered. The details complicate the wave metaphor. Finally, everything about the biology gets too complex and Szilard has to get out of his bath. The next time he enters the water, hoping for a relaxing bath, he remains very still, trying not to set off the cascade of thoughts. Suddenly, "Atchoo!" and his body convulses. Ha, El Niño!

It's a pretty good metaphor, even if it isn't poetic. Fish populations respond to the many processes that influence them in the ocean, kind of like the way that the water level in a bathtub is influenced by waves of different amplitudes and frequencies, measured over different intervals and at different places. Actually a Russian scientist named G. I. Izhevskii beat us to it in 1964 by describing the interaction of environmental cycles in the ocean and fish populations.[3] He described how environmental processes were cyclic with different periods. When he lined them up and added them together he found that the summation of the environmental frequencies matched the changes in abundance of cod in the Barents Sea. Since we've heard almost nothing more of Izhevskii's forecasts for the last fifty years, we assume that they didn't work out very well beyond the time frame he examined.

The geographic basin and wave metaphors are visual aids. By statistical fitting procedures you can take three waves with varying frequencies and adjustable amplitudes and dial up a pretty good comparison to any time series. This demonstrates how attributing biological meaning to complicated models that are fitted to a time series of data is tricky and caution is needed. With that in mind, biological communities and their animal and plant populations do respond to high and low frequency changes in environmental signals that vary in magnitude. They respond over time and also over space. Any one animal population within the community, say pollock, is going to respond to the physical changes, like temperature and currents, and to changes in both predator and prey populations, as well as to parasites and diseases within the community. A scientist sees different patterns depending on where and how frequently measurements are made. Slightly different responses at any level in the system may set off unpredictable patterns. Fish move through space and have individual and collective behaviors. All of this complexity makes forecasting difficult.

One prevalent notion of biology is that we should try to reduce our understanding of biological processes to the basic principles of physics, like adding wave amplitudes. But considering the complexity of biological interaction, it may be preferable to consider physics as the boundary condition of biology, not the basis of it. Physics puts limitations on the potential response. For example, the landscape physically limits the extent of habitat. The energy in the system limits that available to the fish.

Another way to think about biological processes is to focus on an aspect of a problem at different scales from molecular to global. This is known as hierarchical ordering. At any particular level of scale, the problem is influenced by the constraining boundary of processes at the scales higher than the focus, and by the multiple interacting factors at lower scales. In the process of rising and falling fish populations,

| | Lower level | Focal level | Higher level | |
|---|---|---|---|---|
| Noisy factors, stochastic impact | Organism, → patches | **Local populations, stocks** | Ecosystems, metapopulations ← | Constraining factors, deterministic impact |

| Including: | Including: |
|---|---|
| high-frequency environmental effects | low-frequency effects |
| turbulence | long-lived predators |
| prey availability | habitat size |
| advection | physical barriers |
| planktonic and migratory predators | adaptations |
| intracohort density processes | community structure |
| | intercohort density processes |

Figure 7.1 A conceptual view of the recruitment process with high-frequency lower-level factors and constraining lower-frequency higher-level factors influencing the outcome at the focal level of the local population.

there is complexity in the shifting balance of bottom-up and top-down control of the pollock population. In other words, the small-scale factors that control the day-to-day survival of individuals may oppose the big overriding events that may influence the population as a whole. For example, under some conditions, the population may be constrained by higher-order levels, such as temperature and predator abundance—or even very high-order factors like the amount of habitat available. At other times population may be fluctuating from the accumulated interaction of lower-order factors acting on short-term mortality rates of individuals, such as storms, the vagaries of currents, or predation by a flock of birds, and their combined effect on larval and juvenile survival.

## Natural History of Pollock

Let's examine some of the complexities of pollock biology.[4] It's important to point out that this is a relatively new research topic. At the time the pollock fishery was reaching a peak level in 1973, there were only two papers published on pollock. When the MCFA was signed into law in 1976 it was easy to learn new things about pollock as a fisheries biologist. It was uncharted territory. Almost nothing was known about them. In fact, by 1978 there had been only two papers published specifically on pollock ecology by American scientists.[5] Even though pollock is the world's largest harvest food fishery, we still know surprisingly little about them in comparison with other fishes. Every new biological finding adjusts our understanding and reveals how little we really know.

Pollock live mainly in the deeper water over the continental shelf. They gather in large schools and migrate to discrete spawning locations. These spawning areas seem to be consistent from year to year, often in deep bays or canyons and valleys that penetrate the continental shelf. However, there are also offshore spawning groups that spawn over the very deep water of ocean basins. There is some evidence of philopatry, or returning to the location where they spawned in previous years. But we don't know how they navigate the ocean. Extreme philopatry has been observed in pollock's close relative, Atlantic cod, where they are observed to spawn very close to previous spawning locations.[6]

The time of pollock spawning varies somewhat from location to location, but is usually intensely concentrated, and fairly consistent, in any one location. Females release several batches of eggs over a period of a few weeks. In some locations spawning may be as early as January, and in others as late as June.

The role of parenting in pollock is to produce large numbers of offspring and to put them in a location at the best time to increase the odds of their survival. The females also supply the yolk that nourishes the eggs and young larvae through development until they can feed independently. The spawning process begins with seasonal changes in daylight and temperature, which sets off a sequence of developmental changes in adults bringing the eggs to maturation. Eggs have undergone the process of meiosis and divided into the million or so embryos to be released by each mature female that season. A common misunderstanding is that fish like pollock form huge spawning schools of males and females and there is a massive group release of eggs and milt to achieve fertilization. Pollock do form large schools at spawning time, but instead of group sex, individual males and females pair off, and they display an elaborate courtship and mating behavior.

Eggs are generally spawned deep in the water column and, owing to their buoyancy, they begin to slowly rise toward the surface. How deep they are laid depends on the geographical area of spawning. In the

Aleutian Basin the eggs are very deep in the water, at 300 to 400 meters. In Shelikof Strait they are 250 to 200 meters deep, and over the eastern Bering Sea shelf they are shallower, at about 100 to 200 meters. The fertilized eggs are tiny, just about 1 mm in diameter, and translucent. They float freely in the water and gradually ascend toward the surface. The eggs develop for two to three weeks, and then they hatch. The new larvae lack a developed eye, so they can't see, and lie suspended in the water column, sporadically swimming up and then sinking. This vertical ballet results in a net movement of larvae toward the surface layer where their prey are concentrated.

As the larvae develop, their eyes form and become pigmented and functional. Their swimming becomes more horizontally oriented. For the first few days they absorb and gain nutrition from their yolk, their mouths form, and they have just a few days to transition from the yolk to an outside source of food. The transition from yolk to exogenous feeding is a critical time. When they begin to feed, their main prey are small nauplii of copepods, tiny ocean crustaceans, and small invertebrate protozoans like dinoflagellates and rotifers. These prey are not uniformly distributed but are concentrated in patches in the warmer upper layer of the ocean.

The young larvae are fragile and vulnerable to the many dangers they face in the ocean environment, ranging from a vast array of planktonic predators, to starvation from a lack of food, to strong currents that can carry them far from the best nursery areas.

As they learn to feed, larvae change their swimming behavior to maintain themselves in the patches of prey. When they are successful at finding prey they tend to swim slower and turn more, behaviors that will retain them in the prey concentrations. Conversely, when they are not finding prey they tend to swim linearly and faster to maximize the probability of encountering a patch of prey.

Pollock larvae prefer low light conditions, but are also dependent on conditions bright enough to see their prey. Pollock tend to avoid turbulent water. There are complex interactions of the larvae's response to the array of environmental factors that can influence feeding. For example, multiple environmental conditions interact to reduce the survival of larvae during storms. High winds generate turbulence in the upper part of the water column, and cause larvae to dive deeper. This response puts them at depths where light levels are very low, making it harder for them to find prey. On the other hand, when it is windy and bright, both the larvae and their prey are concentrated at deeper levels together but the larvae can now see their prey, resulting in good feeding. Warmer water temperatures allow larvae to feed at a higher rate of ingestion, but their higher metabolic rate causes them to burn more energy. With increasing temperature, there is a crossover point when they burn more energy than they can ingest.

Pollock eggs have a hard cuticle so they are resistant to predation by some small invertebrates. But others can puncture the eggs to get at the yolked embryo. Larvae have poor escape responses for the first few days after hatching and are vulnerable to a large range of predators, from jellyfish to filter-feeding fishes like herring.

At first, eggs and young larvae are very concentrated in the ocean. Sometimes very high concentrations are associated with eddies, or large swirling currents in the surface layers. These eddies tend to move slower than currents moving the waters around them and help to retain larvae near the spawning regions. Pollock have a tendency to spawn in locations where eddies are likely to form.

As the larvae search for prey and also die, they become more diffuse. In the laboratory young larvae don't show social behaviors toward other larvae for several weeks. Then, later-stage larvae begin to recognize each other and orient accordingly in the laboratory. In the field, it is at this stage when the larvae begin to concentrate again, and we recognize it as the beginning of schooling behavior.

Young pollock feed on larger and larger prey as they become capable. Finally, when they reach a size of around 20 mm, or just about an inch, they are completely developed and recognizable as young pollock. They have a full complement of fin rays and ossification of the bones has started, and we call them juveniles. Compared with some other species, which undergo a drastic metamorphosis or change in form from juvenile to adult stages, pollock don't have the drama. Instead, their change from juvenile to adult stages and from a pelagic stage of life to a semi-demersal stage is gradual and perhaps by a series of fits and starts. But by about six months of age, juveniles have moved out of the upper pelagic layer of the ocean and adopted a more demersal stage near bottom.

Studies show that when pollock are confronted with complex problems about their environment, they display sophisticated behavior in making choices. For example, pollock won't migrate from warm to very cold water unless they are motivated by prey being there. They make a decision to risk the numbing effects of extreme cold for a meal.[7] If they stay too long, they go into cold shock. Likewise, in field studies their vertical movements are usually related to a pattern of upward movement at night to feed and downward in daytime, perhaps to avoid predators. However, they can change these patterns to adapt to prey availability.

## Population Dynamics

Since the time of Hjort we have recognized that marine fishes have one or more "critical periods" when they are vulnerable to massive mortality. The period of transition from living on the stored yolk provided by the mother to feeding on prey is believed to be one critical period. Other critical periods may happen when they switch prey types, or change their distribution in the water column. There may be seasonal critical periods as well, such as in autumn when zooplankton abundance is depleted, or in winter when they feed very little and live on stored energy.

Fisheries biologists often look for a dominating process influencing population abundance, or recruitment. Our usual approach is to compare stimulus and response: measurements of changes in the environment and changes in population abundance through correlations. These relationships almost always fail over time for a number of reasons as described below.

Complex processes such as those controlling animal abundance often have unexpected patterns. There are many interacting factors influencing fish populations over a period of years and their relative effect on survival may change over time. Furthermore, the estimation of the factors, as well as the population levels of marine fishes, may be determined imprecisely. Another of the problems is the scale of measurement. We could be measuring a short-term relationship when there is an overlying longer-term trend that we can't see. For example, all animals are sensitive to a range of temperature extremes. A decadal warming period might benefit a species during a longer-scale cold millennium; whereas during a warm epoch, a shorter-term period of warming may be detrimental. Our time series of information on fish populations is too short to recognize these complications. All of these things make predictions difficult.

With all of this complexity in mind, we catch a glimpse of how difficult it is to manage marine fisheries. We think that in many regions of the North Pacific, Alaska pollock increased in abundance in the late 1960s. I say that "we think" because we don't really know for sure. There were few surveys in those early years; those that took place were not designed to sample pollock. The fisheries statistics are unreliable for such a determination because catches were underreported.

Ironically, the event that led to the dominance of pollock in North Pacific coastal ecosystems may have been the destructive effect of overfishing on another species, the Pacific Ocean perch. In ocean ecosystems there are linkages between species that may be strong and direct, or may be less obvious. Pacific Ocean perch are the foundation of a climax community, the equivalent of ancient forests of the sea. They live to be about one hundred years old.

The relentless harvesting of the Pacific Ocean perch by Soviet and Japanese fisheries in the 1960s and the depletion of the population was accompanied by the growth of the pollock population. Pollock probably responded to less competition and newly available prey resources, and possibly less predation by perch on young pollock. It was like clear-cutting an old-growth forest, allowing the undergrowth to blossom and to set up a new succession of community. Pollock exploded onto the scene. A favorable climate shift in the North Pacific Ocean during the late 1970s was responsible for a further abundance spike owing to increased birth and survival rates. Being opportunists, pollock took advantage of the situation.

The exploitation of pollock itself may be a factor involved in other ecosystem changes. One ecosystem change has been an increasing abundance of jellyfish in the Bering Sea since the mid-1970s. It has been proposed that the pollock fishery reduced the abundance of the pollock preying on zooplankton, releasing the zooplankton as prey to jellyfish, thus fueling their growth spurt.[8] Broad ecosystem linkages of commercial fishing, shortages of prey for sea lions, killer whale diets, declines of sea otters, and kelp/sea urchin dynamics have also been made.[9]

One of the difficulties of ecology is that answers obtained by observation are rarely definitive. There are usually other explanations.

Another plausible explanation for the increase in pollock is that there was a change in some other environmental condition that was disadvantageous to the Pacific Ocean perch, while it was favorable to the proliferation of pollock. Likewise, jellyfish may have increased owing to a change in some other unknown factor. The issue of the sequential megafaunal collapse and relationships of commercial fishing with sea otters has been contentious.[10] Usually ecologists have to sort through the alternative hypotheses and find out which seems most reasonably supported by the available data.

The recruitment of pollock in both the Bering Sea and Gulf of Alaska varies by 100-fold from year to year. The amount of year-to-year variability caused by environmental effects on survival rates of young pollock obscures the relationship of recruitment with the size of the parent population. For both of these populations, strong year classes are essential to the health of the overall stock. In the Bering Sea there is no clear trend in the time series of the recruitment level except in recent years, when there has been a downward trend. In the

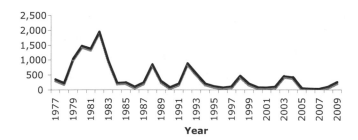

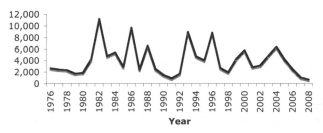

Figure 7.2. Recruitment levels of age four pollock in each year as it entered the fishery.

Gulf of Alaska there has been a downward trend since the early 1980s and the height of the peaks has declined.

The harvest of the Gulf population has been taken at a much higher rate compared with the eastern Bering Sea. An interesting aspect of the Gulf of Alaska population is that they seem to have many more spawning areas. The habitat is along the continental shelf, which is narrow and elongated, and fragmented by numerous deep sea valleys which cut across it. The pattern of exploitation has been for the fishery to fish one area where there is an aggregation until it is depleted and then move to another area. By contrast, in the eastern Bering Sea the overall population is much larger and the main spawning over the shelf is concentrated in two broad regions. The shelf habitat is broad and expansive.

One notable difference between the regions is the degree of cannibalism observed among the populations. In the Gulf of Alaska, extensive cannibalism has not been observed and appears to be a much less prevalent feature of the population. On the other hand, in the eastern Bering Sea, cannibalism is extensive. There has been speculation that the fishery has allowed increased survival of juveniles due to the reduction of cannibalistic adults. This observation, along with lower harvest levels, is a potential reason for the greater stability of the eastern Bering Sea shelf population. The differences in the landscapes of the two regions and the natural histories of the species contained in them underlie the different behaviors. The depleted populations in the Aleutian Islands and Bogoslof Island were also harvested at high levels and were not known to be extensively cannibalistic.

Another important factor affecting recruitment levels in the Gulf of Alaska is the high abundance of another predator of juvenile pollock, the arrowtooth flounder. Beginning in the early 1990s, arrowtooth flounder surpassed pollock as the major groundfish species in biomass in the Gulf. At the same time, increases in juvenile mortality were observed and rates of larval mortality became decoupled from recruitment levels. Predation by flounder may be constraining the survival of juvenile pollock. The arrowtooth flounder abundance has increased in the Bering Sea, but not nearly as much, and pollock are still the overwhelmingly dominant species there.

Although predicting pollock recruitment has been a hard nut to crack, we have made great strides in measuring how larvae respond to environmental conditions. An individual female spawns millions of eggs over her lifetime and on average only two need to survive and reproduce for replacement. With so many larvae produced each year, the role of larval survival is believed to be important in production because a small percentage change in their mortality rate, which is very high (10% to 30% per day), can have enormous consequences to later numbers. The knowledge gains made in recent years have been particularly strong in assessing the growth, condition, and mortality during the larval stage.

The condition of larvae can be assessed in a number of ways, from examination of individual tissue condition to measurement of critical storage or growth components in the body, or even by estimation of cell division rates using lasers coupled with nucleic acid staining dyes. The results indicate that high percentages of pollock larvae can be found in starving condition at certain times and locations.

The actual growth rates of larvae can be determined quite precisely from stones found in the inner ear of fish, called otoliths. The otoliths are used for balance by the fish and determining inertia. The otoliths lay down annual increments similar to the way trees lay down rings. But if you prepare them carefully and view them under a powerful microscope, increments that are laid down daily can be observed. When one knows the age and size of a fish, the average growth rate over the age interval can be estimated. Daily growth rates can also be estimated from the distance between the increments.

If the age is known, and abundance is also known, then we can monitor the decrease of animals of a specific age group over time. In other words, we can get quite precise estimates of larval mortality. Using these techniques, the growth and mortality can be compared with factors like temperature and prey availability. In the case of pollock, both temperature and prey availability are important factors in growth and mortality.

Predation on eggs and young larvae is believed to be important in their survival. The number and diversity of predators is overwhelming, from predatory copepods, amphipods, krill, shrimp, and jellyfish to pelagic fishes. The interaction of predation with larval condition, and the ability of larvae to escape predators, are complexities of the recruitment process.

The problem in predicting recruitment, then, comes from putting

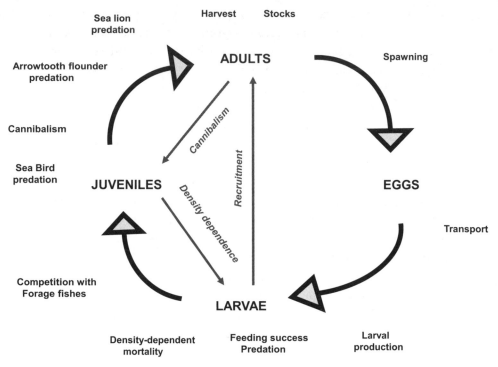

Figure 7.3. The complex life cycle of pollock with some of the dominant factors occurring at each stage that influence recruitment to the adult population.

together the bewildering number of external factors influencing survival. There are also intrinsic factors in the population; cycles that occur as a result of interactions between age groups, such as cannibalism; as well as interactions of the intrinsic factors with the external ones. Looking at the time series of recruitment, it appears that there is some periodicity in high and low recruitment years. For the Gulf of Alaska population, it is a statistically significant cycle. However, there is no strong solid understanding of the cause of the cycling. It is tempting to compare one or two environmental factors with recruitment and to be absolutely convinced that you have found the final solution. It may indeed be a part of the answer for the moment, but almost always, the simplistic answers fail over time.

That being said, animal populations are regulated by certain natural laws. Their abundance and distribution are crudely determined by their adaptation to environmental parameters, such as temperature.

It is well known that certain trends exist such as abundances declining at the warm end of a species range when temperatures increase, and vice versa at the cold end of the range.

## Population Structuring

The architecture of populations, or the distribution of populations in space, is an important characteristic of natural fish populations and should play a critical role in how they are managed. But it is a subject area where scientists are exploring a relatively new frontier. Populations that are separated by barriers or long distances, or that have remained apart for long periods since interbreeding, have the opportunity to differentiate themselves genetically and also behaviorally. The rate of differentiation depends on factors such as the migration rates among populations and population size. Behavioral changes, which include the timing of spawning and mating rituals, reinforce the barriers to interbreeding. Local populations also adapt to local environmental conditions through selection.

The architecture of fish populations, sometimes called the "biocomplexity," affects the aggregate population's resilience to local extinctions and reflects its tendency to colonize marginal areas. The role of landscape, the geographic complexity of suitable habitat, and oceanography are key aspects of fish population structure that are just now being recognized. Development of technology is giving us new tools to examine genetic population structure and migrations at finer levels than before.

Relatively few resources have been devoted to studying the structure of the Alaska pollock. Past and current studies indicate that there appears to be a strong separation of populations across the Pacific Ocean separating western Bering Sea and Japan populations from those in the eastern Pacific. There also appear to be some minor differences among the major eastern Pacific populations. Given the minor differences, much larger sample sizes are required than have been used in the past to reduce the variation found in small random samples. Nevertheless, the studies to date point to minimal exchanges of genetic material among the major population areas—for example, western and eastern Pacific, and Puget Sound and Bering Sea.

The close relative of pollock, the Atlantic cod, has been more thor-

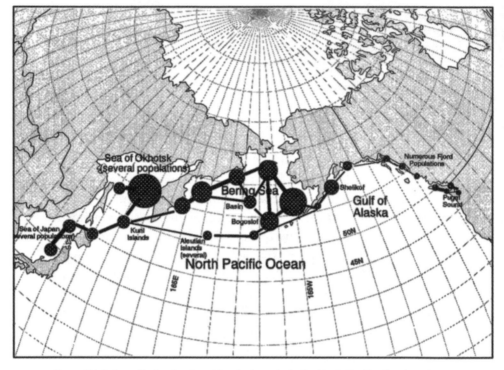

Figure 7.4. Schematic for structure of pollock stocks in the North Pacific. The size of circles reflects the relative population size and the width of lines represents the putative relative exchanges.

oughly studied. In the Atlantic cod population, much more structuring of populations on very small scales has been found. Tagging studies of cod have indicated fidelity to spawning areas on very small scales. These intriguing studies are important to consider for management, and point to a need for a more thorough investigation of pollock stocks.

The structure of populations is important to consider in establishing where, when, and how many fish can be harvested from a region. Local populations can be fished to extinction readily, especially when they are close to a home port. If there is deep structure, or little movement between local groups of fish, then recolonization of the depleted population may take a long time. This diminishes the overall productivity of fish in the region. If there is a lot of mixing among local populations, then local fish depletions won't have a long-lasting impact on the health of the overall population.

Knowing the details of how fish migrate is particularly important in management. It is known that pollock range widely in search of food in the summer feeding period, but they return to recognized spawning areas year after year. The various mechanisms they use to accomplish their successful return may be critical to harvesting strategies. For example, if there is a social component such that new recruits follow older and more experienced adults to spawning locations, then harvesting the larger fish can have severe implications. One imagines that as the older fish become depleted, the probability of younger fish encountering them declines. Then the young are more likely to start spawning in a new location, which may be less productive for survival of their young. In doing so, they attract more spawners, and a temporary spawning trap develops until the population spirals downward. Most likely, fish use an array of clues when they return to their spawning areas, including imprinting on natal locations and a genetic component based on local adaptation. One could speculate that there may even be an epigenetic effect (an inherited phenotypic change of gene expression), unrelated to a change in DNA sequence.

### Recent Stock Dynamics

The eastern Bering Sea shelf and northwestern Bering Sea populations seem to be remarkably steady over the past 40 years compared with many other fish stocks. The eastern Bering Sea group underwent a decline over the past ten years, causing quotas to be reduced. The fishery has been under close scrutiny from managers, environmental groups, and fishermen in recent years.

The Donut Hole stock, which was discussed extensively in Chapter 5, hasn't been the only population to suffer devastating depletions. The declines in the Bogoslof and Aleutian Islands stocks were also discussed. Western Bering Sea pollock on the eastern side of Kamchatka were severely depleted in the early 1980s, going from a biomass of 7 million tons in 1984 to around 4 million tons in 1990. It leveled off and then began another slide in 1992 to less than 1 million tons in 2000.[11] In the early 1990s, part of the American Seafoods' fleet under

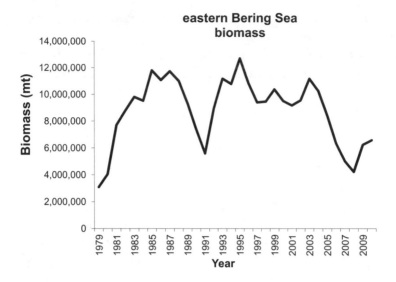

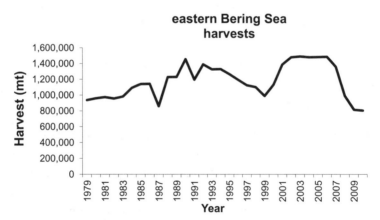

Figure 7.5. The estimate of the stock biomass of fish three years and older (top panel) and the harvest of pollock on the eastern Bering Sea shelf (bottom panel). Note the high catch levels in 2004–2007 when the stock was declining. Although the biomass is age three and older, most of the catch is four years and older.

Kjell Inge Røkke harvested pollock in Russian waters because of the overcapitalized fleet and the shortened fishing season in US waters. American Seafoods entered into a partnership with a Russian company in Vladivostok to fish the western Bering Sea off the coast of Kamchatka with several Spanish-built super trawlers.[12] These ships had been paid by the European Union to cease fishing off Europe.[13] Several other American and foreign companies followed American Seafoods to fish in the western Bering side. There are reports of large bycatches and discards of small pre-recruit pollock.[14] That population experienced steep declines in the early 1990s, and harvests of the western Bering Sea stock of pollock are now less than 1% of previous levels. However, there has been some evidence of a recent modest rebound.[15]

There was also uncontrolled fishing in the international waters of the Sea of Okhotsk.[16] The stock biomass declined severely, from about

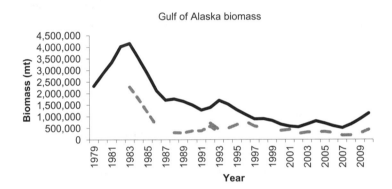

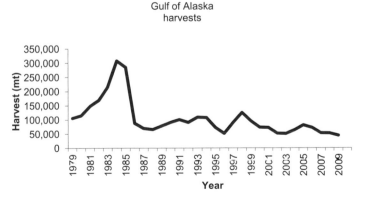

Figure 7.6. The total biomass of the stock in the Gulf of Alaska greater than three years of age (top panel solid black line), the total biomass of pollock in the Shelikof Strait only (top dashed grey line), and the harvest of pollock in the Gulf of Alaska (bottom panel). Note the peak catch occurs as the stock is already declining. Most of the fishery in the 1980s was in Shelikof on adult spawners.

10 million tons in 1995 to less than 2 million tons in 1999, but has been slowly rebuilding.[17] There was fear that the stock would be fished to extinction by Japanese, Vietnamese, Korean, Polish, and Russian fishermen. Catches were vastly underreported.[18] However, recently the Sea of Okhotsk population seems to be holding steady.[19] There were allegations of involvement of the Russian mafia in this fishery, and corruption in Russian fisheries appears to be an ongoing problem.[20] Illegal plundering of pollock in the Sea of Okhotsk continues now by the Russian fleet with China as the major buyer.[21]

The Food and Agriculture Organization (FAO) of the United Nations reports that pollock stocks in the Kuril Islands, off western Japan, and in the Japan Sea have been heavily fished and biomass levels there are substantially lower than in the 1980s.[22] Stocks in 2008 were reported to be at their lowest levels in twenty-eight years.[23] There is a report that one spawning ground of pollock in the Sea of Japan has disappeared.[24]

The population in the Puget Sound experienced a serious decline in the late 1980s. It appeared to be practically extinct in the late 1990s and was reviewed for possible listing as an endangered species. Another major population in the Gulf of Alaska has tottered on the edge of collapse. Shelikof Strait used to be the dominant population in the Gulf of Alaska, but now the estimate of the spawning stock biomass in Shelikof is only 7% of its maximum level in 1981. After the fish in Shelikof declined, the fishery moved to the Shumagin Island and other regions to harvest pollock. There is a cut-off for the fishery that occurs when the population falls below 20% of the maximum. Since the North Pacific Fishery Management Council (NPFMC) lumps these smaller local populations in the greater Gulf of Alaska population, the total spawning biomass has been kept above that critical value.

# 8 A New Ocean: Changing Concepts of Ocean Production and Management of Fisheries

In the 1970s the models used in fisheries were largely unquestioned. The fisheries scientists Hjort, Graham, Beverton, Holt, Ricker, and Schaeffer had ascended to Olympian stature in the world of fisheries management. There was a strong tradition, passed from professor to student, and an air of certainty that fisheries science was on the right track.[1] Most fisheries scientists came from biology or zoology programs and the concepts of fisheries management were new and mysterious, founded on the application of mathematical models to populations. There is something about describing populations with models that makes it seem as if we understand them better than we actually do. Scientific paradigms usually change more slowly than we'd like to think. It often takes the passing of the old generation before new ideas become mainstream.

The status quo has been challenged recently. Some of the students from the 1970s were activists who questioned authority. Some asserted that, rather than being an exact science, "Population modeling is an inexact science. Populations are part of complex systems that defy the taming tethers of 'physical laws.'"[2] Referring to the old-school use of fisheries management models, one writer said that we had put the cart before the horse. Models should be used to infer something about nature, rather than the other way around![3] The ocean's resources began to look more limited. Fisheries were being depleted in spite of management by maximum sustainable yield (MSY) criteria. A new group of scientists and environmentalists from outside the traditional aca-

demic fisheries lineage, as well as some from within, such as Sidney
Holt and Alan Longhurst, started rocking the boat.

Several major events signaled the new era in the world fisheries
view. Policy actions, such as the FCMA of 1976, were turning points
in American fisheries, extending the fisheries conservation zone to
200 miles. This trend expanded worldwide with the Law of the Sea in
1980. Sovereign nations now had more control over coastal resources.
The Endangered Species Act of 1973 gave federal agencies new power
to protect threatened species. The Sustainable Fisheries Act of 1996
shifted the emphasis from fisheries development to conservation. Natural events such as El Niño, its effects on marine systems, and the collapse of Peruvian anchoveta, focused news reports on the ocean, usually because of the severe economic consequences. Largely manmade
disasters, such as the collapse of California sardine and northern cod,
pointed out the complexities of human-fisheries-environment interactions. These events were eye opening. Perhaps we couldn't control
nature as well as we thought.

Concern for the future of the ocean environment was popularized
by television programs, such as "Flipper" (1964–67), "The Undersea
World of Jacques Cousteau" (1966–76), and "Nova" and "National Geographic Specials" that started in the 1970s. Ocean advocates turned the
plight of marine mammals into emotional and popular public topics.
Whereas in previous decades we were not aware of what was happening in the ocean, it was now in our living rooms.

As a result, how we view the oceans has turned upside down in the
past forty years. Although scientists studying individual stocks had
recognized that fisheries depleted them, the popular concept of a
boundless ocean had prevailed. But that perception began to change.
In 1993 the worldwide harvest of fish exceeded 100 million tons for
the first time.[4] Currently the world's harvest of all capture fisheries
has leveled off at 90 million tons (and that of marine fishes at about
70 million tons), and that level is arguably too high. The ocean was
considered unlimited, but gradually we realized that there may be
some limit, and now we are backing off. Globally we may have surpassed sustainable levels some time ago.

Still, the view of a boundless ocean has hung around like a low-
lying fog. It generally takes the government about ten to twenty
years to catch up. As Michael Weber noted, "Until the 1990s [US] fed-

eral policy and practice were generally based on the belief that the ocean's productivity was almost limitless and could be manipulated for maximum production and utilization. The chief goal of policy was to increase the capacity of US fishing fleets to exploit this abundance." Around the same time many scientists and politicians and the public began to realize that fisheries were not unlimited. In the 1990s the theme shifted from *abundance* to *scarcity*.[5] Suddenly we began to hear that fisheries were overcapitalized and overexploited.

Fisheries managers have been vexed with the same two questions since the beginning of their profession. How many fish are in the sea, and how many of them can we catch? The methods of counting fish and assessing populations have improved since the early twentieth century. A lot of effort has gone into standardization and calibration of trawl surveys. The use of acoustics to measure fish abundance is more prevalent and better tuned for improved assessments. The models developed to assess fisheries abundance from catch statistics in the 1970s have become more sophisticated. But regardless of improvements in assessment methodology, there is still great uncertainty about the estimates of fish abundance. Most problematic, fish are often tightly clustered and easily missed by surveys, they move, and they often spill over the boundaries of surveyed regions.

Even if we have good estimates of fish abundance, then there is the question of how many can be harvested. Prior to the 1980s, a sustainable fishery was defined as one where the fishing pressure was regulated so the population level could be maintained. The FCMA of 1976 required NMFS to determine in which fisheries the rate of exploitation jeopardizes the capacity of the fishery to produce the maximum sustainable yield on a continuous basis. That was later modified to include economic sustainability (i.e., profitability).

During the growth phase of American fisheries, up until the mid-1980s, scientists in the United States were under the pressure of open access to fisheries and the need for expansion.[6] The concept of MSY encouraged fisheries managers to set quotas to fish the population to a lower population level in the belief that this would allow younger fish to grow faster and would boost productivity.

According to Longhurst, the central axiom of fisheries science is that fisheries are sustained by the density-dependent increase in growth that is provoked by a fishery, the surplus production.[7] The basis of MSY, the logistic curve of population growth, describes changing birth and death rates with increases in population under the assumption that growth slows at a maximum size under the conditions of limited resources. The population production is maximized, or optimal, at about 50% of the maximum population size. The concept seems logical enough, but it is hard to find circumstances where data reliably support it. Harvesting isn't the only event in the arena. Food supply changes with time, predation is varying and also adjusts the abundance of prey species, and social behavior changes with density. The fisheries models seemed elegant in their simplicity, but they are very rough caricatures of nature.[8]

In the 1960s the question of the certainty of density-dependent versus independent population growth was a contentious issue in ecology. Many ecologists reached a compromise when they decided that both were important. Circumstances may change where one or the other becomes a dominant factor in population dynamics. These debates have been largely ignored by fisheries scientists through the assumptions of logistic population growth and density dependence, and the concept of optimal yields. Longhurst noted that most fisheries continue to be managed as if they existed in a perfect invariant ocean. The idea that biomass and catches are ultimately driven by climate fluctuations runs counter to the precepts of fisheries management, which assumes that biomass and catches are driven by fishing pressure.

When cod stocks in Canada declined, the models suggested a density-dependent rebound. Instead, Longhurst reported, "Canadian biologists at the Department of Fisheries and Oceans were said to have been stupefied when it became clear that predicted strong growth of the northern cod stocks had not, in fact, occurred. . . . Their confidence in simple constructs appears to have been nurtured by assumptions that the universe is mechanistic and governed by a few simple laws . . . and that the state of the marine ecosystem varies around a dynamic equilibrium."[9]

One problem with current management concepts is that declining populations don't necessarily behave like healthy ones. They may be less productive, often owing to environmental conditions, and are under extra stress owing to something called *depensation*.[10] When fish populations are at low levels they don't always follow the classic logistic form of growth that is one of the assumptions of surplus production models; that is, they may not be more productive at lower densities and with more resources available. Many reasons have been put forward for depensation. One explanation is predation. When a population is at a high level, predators increase and their feeding areas shift to take advantage of localized increases in prey. When the population decreases, the predators are still there and there is a lag in the behavioral response to disperse and find other prey. The result is strong ecosystem-generated mortality levels. Combined with even moderate fishing mortality, the enhanced mortality due to predation can cause an accelerating loss to the population. This form of depensation probably contributes to uncontrollable collapses of marine fish stocks.

Competition is another cause of depensation. When fish populations are reduced to low levels, the idea of MSY is that they will grow faster because there will be more food available to the remaining fish. However, competing and less desirable species may also take advantage of increased prey availability. In this case, the effect of selectively harvesting a population may be like mowing the lawn with a dull-bladed machine: the thistles that are resistant to the blades may prosper most from the thinning by reduced competition.

### Shifting Baselines

In the northeastern Pacific Ocean, yellowfin sole, herring, and Pacific Ocean perch were depleted by the distant-water foreign fleets in the 1960s and 1970s.[11] Globally it appears that the view of an unbounded ocean resource was collapsing. A report filed by the United Nations in 1990 indicated that nearly every commercial species surveyed was either fully exploited or overexploited, or even collapsed. A fervor to document globally collapsing fisheries gained urgency that peaked in 2006 with the report by Worm et al. that in 2003 about 30% of the world's fish stocks had collapsed, and the projection that there would be 100% collapse by 2048.[12]

The depletion of top-of-the-food-chain predators was the most dramatic, and it was argued that 80% to 90% of the original biomass top predators had been removed by fisheries by 1980.[13] The consequence to marine ecosystems would be a remarkable reshuffling, perhaps resulting in the domination of many communities by jellyfishes. The fishing industry was now "fishing down" the marine food chain to exploit animals that are more abundant closer to the base pyramid structure.

Bottom trawling in particular was reported to be destroying bottom habitat in large regions of the ocean, upsetting the benthic habitat and reducing the productive capacity of the oceans.[14] Loss of biodiversity due to overfishing and consequent loss of system productivity was another issue.[15]

Another camp came together, which has rebutted the pessimistic views of the state of marine harvest fisheries, saying that fisheries management is doing well in managing stocks at MSY for many cases.[16] They reported that estimates of overfished and depleted stocks have been exaggerated.[17]

One of the most outspoken proponents of fishing stocks according to the principle of maximum sustainable yield (MSY) is Ray Hilborn, professor of aquatic and fisheries science at the University of Washington.[18] In a 2011 review of Longhurst's book *Mismanagement of Marine Fisheries*, Hilborn said, "Longhurst would be the first to point out that many stocks now classified as 'overfished' are at low abundance largely because of changes in ocean productivity, not because of fishing."[19] However, Longhurst indicated that this statement misrepresented his view. "I don't know where he got that idea from."[20] There is another nuanced interpretation of what Longhurst actually writes in his book. Longhurst points out that the central axiom of fisheries science is that fisheries are sustained by density-dependent increases in growth rates provoked by fishing, the so-called surplus production. The point of his book is that the basic tenet of fisheries management, the concept that they can be managed on the basis of an estimate of surplus production, in the face of ecosystem complexity and natural variability, needs to be scrutinized more carefully. Fisheries are inherently unsustainable entities, rather than systems that can be engineered and regulated.

Hilborn is one of the best known and highly regarded fisheries scientists. He often speaks out on a wide variety of issues. He was reported to say in October 2011 at a meeting sponsored by the fishing industry and appropriately held in Fishmonger's Hall in London, that the yield of fisheries by the major industrialized nations could be increased 41% by fishing harder on underexploited species, which sounds eerily similar to Rachel Carson's advice during World War II. In response to the antifishing group's warning that the ocean will have nothing left to eat but jellyfish, Hilborn said, "We shall have nothing to eat but bluefin tuna," presumably tongue-in-cheek.[21]

For most of us with even a modest amount of expertise in fisheries, it is hard to sort out which of these analyses is correct or whether they merit serious consideration. There are many ways to define "over-fished" and "collapsed," a lot of different data sources to sift through to support a particular viewpoint, and different ways to analyze the data. As with most issues, the truth probably lays in between the extreme viewpoints. Given the history of NMFS in promoting fisheries development, one could hardly say that the government's official position on fisheries issues is objective either. Another prominent fisheries scientist, Daniel Pauly at the University of British Columbia, wrote, "Our oceans have been the victims of a giant Ponzi scheme . . . carried out by nothing less than a fishing-industrial complex—an alliance of corporate fishing fleets, lobbyists, parliamentary representatives, and fisheries economists. . . . They secured political influence and government subsidies."[22]

One fisheries conservation group claims that the NMFS has been guilty of pulling the wool over our eyes by lumping together healthy and declining stocks.[23] For example, if one considers only the Donut Hole stock of pollock, it looks like it has collapsed; but if it is combined with all the other stocks in the eastern Bering Sea, the result gets averaged out. On the other hand, some scientists make the opposite claim that exaggerations of overfishing have resulted from lumping different species together.[24] There are the lumpers and the splitters and they come up with different views of the world.

These debates among the profishing and antifishing groups sound similar in tone to arguments over the human impact on climate change or effects of tobacco on health. The average citizen who has a hard time sorting facts from statistical artifacts, or sorting out the different interpretations, may be influenced by advertising. Unfortunately, the winner of public opinion and government action is often the side with the most funds for public relations. Natural resource management is on shaky ground when the science becomes driven by agendas and finances, whether that is originating from industry or conservation organizations, or driven by the pressure lobbyists exert on government.

## Early Pollock Management

The management history of pollock stocks in US waters has been different from that of many other American fish stocks. The early stages of development of the halibut and salmon fisheries on the West Coast left a strong history of regulation through limits on removals, areas closed to fishing, and control of fishnet mesh sizes. From nearly the beginning of the development of North Pacific fisheries, there was a consensus to research and understand fish populations, and generally a strong scientific basis for making decisions.[25]

However, when the pollock fishery developed as a foreign-based industry off the coast of Alaska, and right up until the mid-1970s, there was literally no management of the resource. This is because American fishermen and scientists were not interested in pollock. Then upper limits were placed on groundfish catches from the eastern Bering Sea through bilateral treaties, but these were self enforced and not monitored. Since the United States lacked jurisdiction over fishing outside the twelve-mile US contiguous fishing zone, those foreign fishing nations that agreed to adhere to treaty provisions were rewarded with special privileges, such as permission to fish in certain areas, or the ability to provision or to load ships within the twelve-mile US zone.[26] There were various area and seasonal closures to protect halibut and crab stocks, and to minimize conflicts with American fishing operations. In the mid-1970s, American observers were put on foreign boats to monitor catches through a voluntary program.

After passage of the FCMA extending the exclusive American fishing zone to 200 miles, foreign ships were allowed to fish in American waters by permit, but under stricter regulations. Regulating a foreign fishery was easier than dealing with domestic fishermen, and this established a strong tradition of observing and monitoring catches. This precedent of regulating fisheries is different from the situation that developed on the East and West Coasts of the contiguous United States, where fishermen have grown to be independent and anti–regulation-minded.

Under the FCMA separate management plans were developed for the Gulf of Alaska and Bering Sea/Aleutian Islands. Frameworks were established for the setting of annual harvest quotas, allocation of quotas among domestic and foreign fisheries, area and seasonal closures, gear restrictions, and bycatch limits. An optimum yield (OY) was established for the total groundfish complex in the eastern Bering Sea. The OY is required to be less than the sum of the MSY for individual species in the complex. Then a total allowable catch (TAC) is set as the harvest quota for each species, and the sum of the TAC cannot surpass the OY. In the Bering Sea, the OY of all groundfish removals was set at 2 million tons. A planning team evaluates the stock assessment survey data and modeling results, and recommends an allowable biological catch (ABC), which is generally the MSY when stocks are in good condition. When a stock appears to be declining, extra safeguards are implemented. The NPFMC determines the annual quota starting with the ABC and considering any other factors, such as ecological or socioeconomic concerns, and sets the TAC.

Referring to North Pacific fisheries management, NPFMC member and industry chieftain Wally Pereyra said, "Everyone's talked about the science-based management system we have, and that certainly is a strong part of our success. But the other part of the success is the fact that we have been able to enforce the quotas that we have set, and enforce them to a level of precision that has allowed us not to be overfishing the resource."[27]

Pollock stocks have been monitored by trawl surveys since about 1976 and with consistent methodology since 1982. Acoustic surveys were added in the 1980s. The surveys were initially used to set abundance levels and monitor trends. Later, models employing catch statistics supplanted the surveys as the primary stock assessment method, because the survey data usually gave a low result. The survey data were used to "tune" the models, monitor trends, and independently verify the stock condition.

The fisheries management plan now in place for Alaska pollock uses a complex model that estimates 762 different parameters of the population, and management is a complex interaction of limits and harvest cut-off points. The survey data are incorporated in the model, meaning there is no independent check of the model's assessment. The integrated model result is now considered the "gold standard."

The gold label doesn't mean that the model is perfect. It is still evolving. Thus far, there is nothing in the model that incorporates environmental change, the spatial structure of the population, or the ecosystem, although that work is in progress. There is a chapter in the fisheries management plan called "Ecosystem Considerations" that summarizes the state of the ecosystem, but it is not directly connected to management actions. The model also does not consider the resilience of the population due to changes in the age structure.

Although the population in the eastern Bering Sea hasn't collapsed, we have dodged the bullet a few times.[28] There were dramatic stock declines from 1987 to 1991 and then again from 2003 to 2009. Several scientists are forecasting lower stock abundance in the future due to climate change.[29]

[9.] An American catcher vessel making a delivery to a Korean factory processor in rough weather during in the early 1980s (FMA/AFSC).

[10.] American captain transferring ships in a wet suit. This method was used when seas were too rough to launch a skiff. Generally the man in the survival suit would tie himself off to a buoy, throw it in, and jump in after it (FMA/AFSC).

[11a.] The *Golden Alaska* in the ice, Kjell Inge Røkke's first surimi processing ship. This is the rebuilt ship after it burned to the waterline (from Aaron Barnett, Golden Alaska Seafoods).

[11b.] American Seafood's *American Triumph*. The ship was rebuilt in Norway from the 74-foot *Acona*, and was the subject of Senator Ted Stevens' wrath in Senate hearings (from Jan Jacobs, American Seafoods).

[11c.] The *Arctic Storm* (from FMA/AFSC).

[11d.] The *Northern Jaeger* in sheltered waters of Alaska. Tim Thomas is the captain (from Jan Jacobs, American Seafoods).

[12.] Below deck in the factory on the *Northern Glacier* (FMA/AFSC).

[13.] The Unisea factory, Unalaska (FMA/AFSC).

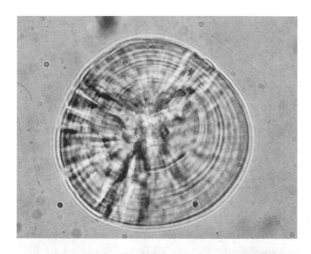

[14a.] An ear stone, or otolith, of a larval Alaska pollock showing daily growth increments (from Annette Dougherty, RACE/AFSC).

[14b.] An otolith of a larval pollock that had been immersed every seven days in a bath of Alizarin, an orange fluorescent dye, showing the weekly bright marks and the six daily increments in between (from Annette Dougherty, RACE/AFSC).

[15.] The head of a larval pollock of about 6 mm length. Just behind the eye in the bottom of a large clear area are the otoliths (from Annette Dougherty, RACE/AFSC).

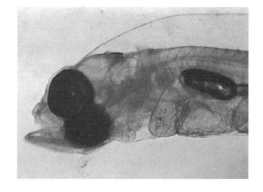

[16a.] Seven Greenpeace protestors hanging from the Aurora Bridge in 1997 (with permission from photographer Larry Murphy of Zendog Studios).

[16b.] A Greenpeace protestor hanging from the Aurora Bridge in 1997 with a sign "Ban Factory Trawlers" (with permission from Larry Murphy of Zendog Studios).

[17.] The community of Unalaska on an autumn day. Cars in the foreground are driving the road across the island to Dutch Harbor. In the background, processing plants are located along the shoreline (with permission from photographer Linda Lowry, provided by Elizabeth Masoni of Unalaska).

[18.] The infamous Elbow Room of Dutch Harbor, home to the hijinks of many fishermen (FMA/AFSC).

[19.] Chipping ice from the hull in the eastern Bering Sea (FMA/AFSC).

# 9 Factories of Doom: The Pollock Fishing Industry Clashes with the Environment

*As fish are caught in a cruel net, or birds are taken in a snare, so people are entrapped by evil times that fall unexpectedly upon them.*                    Ecclesiastes 9:11–13

### Greenpeace Wages War

In August 1997 seven climbers with ropes prepared to rappel down from the Aurora Bridge where Highway 99 passes over the ship canal in Seattle. The bridge is the most popular suicide venue in town. The railings are low and the water is 175 feet beneath. As a jumper's bridge, it ranks number two in the United States behind the Golden Gate. In summer 2011 a safety fence was installed to end the specter of people falling from the sky, especially now that several upscale businesses like Adobe Systems and Google are located on the land underneath that borders the canal. Sometimes the victims came to a final rest in the parking lot if they calculated wrong or didn't anticipate the wind drift.

Ships moored in the lake pass through the Lake Washington Ship Canal and under the bridge on their way to the ocean. At the moment our climbers were about to heave themselves over the railing, a policeman drove past. He noticed the people, commotion, and ropes, and pulled a quick U-turn. With the climbers preparing to descend, he pulled out a knife and appeared to saw the ropes. "I've cut your lifeline, I've cut your lifeline," he shouted. After a few intense moments, a Greenpeace support person looked over the lines and yelled out, "You're safe!" The climbers leaped over the rail and descended.

The climbers included Greenpeace volunteers and activists. At least two of them were associated with the Ruckus Society, a group of environmental activists trained in the art and practice of civil disobedience. A factory trawler approached the suspended bodies. They were all strung together by another line, so if the ship tried to pass between them, it would entangle all and send them plummeting. One climber said, "I've never seen a ship this large before, and that close. And I was toward the middle, so I got a good view."[1]

The ship and the climbers faced off. The captain of the American Seafoods factory trawler searched for a clear path through, then he stood down and returned to the pier. The climbers hung from the bridge for forty-eight hours while national and international television news programs reported the event.

After they climbed back up, a Greenpeace spokesman quipped, "We were able to block two of the factory trawlers. More importantly, we communicated the damage they are doing to the ocean." A representative of the fishing company countered, "It is universally agreed that it's a very conservatively managed process [fishery]. . . . It doesn't do anybody any good when they're simplifying [the issue] into sound bites."[2] One of the fishing boat's crew said, "I understand they're doing it for what they think is a good cause and everything. It's something they believe in. It's just that this is our bread and butter, and it has an effect on us."

The Seattle police arrested the climbers and their supporting staff. They were freed on bail, and finally acquitted in a trial by jury in June 1998. The court found they had constituted a legal assembly. One of the climbers related the support they got from the Seattle community. People brought them hot coffee and snacks while they were suspended in air. One kayaker paddled out on the lake at night with a guitar and serenaded them to sleep. Replying to the comment, "You could have been killed," one activist said, "I don't think about that. I think about what has to be done, and sometimes this is what you need to do. And the being killed aspect just doesn't seem part of it. And there's no choice; it has to be done."[3]

Paul MacGregor is a knowledgeable and respected Seattle attorney who was in the midst of the battlefield. His round eyeglasses give him a slightly owlish look, which is reinforced by a stature of composure and alertness. One gets the sense that he is watchful of his surroundings. When I interviewed him he wore neatly pressed slacks and a sweater vest over a darker shirt.

MacGregor is the chief counsel representing the At-Sea Processors

Association, an industry association of companies that own and operate the large factory trawlers in the Bering Sea pollock fishery. In his position as chief counsel, he knows what is going on in the industry and guards the interests of his clients. MacGregor graduated from Stanford Law School and obtained a master's degree from the London School of Economics. He has worked on North Pacific fisheries issues since 1976. When you talk with MacGregor, you realize that he is a man deeply involved in his work. He represents his clients well. His influence in the pollock industry often gets done behind the scenes, but it is pervasive.

One day in early 2011, MacGregor walked into my office. Coincidentally I had just gotten off the phone talking with Ken Stump. Stump was a highly visible and outspoken Greenpeace activist of the late 1990s at the height of their anti-factory trawler movement. Even though over a decade passed since those turbulent times, when I told MacGregor I had been talking with Stump, he looked furious as he launched into a tirade about Greenpeace. From MacGregor's reaction, I realized just how hot the emotions ran, and they still smoldered. This put me in an awkward position since I enjoyed my conversations with both MacGregor and Stump. They each presented a point of view convincingly and with the passion of true believers. I felt caught between a set of gnashing teeth.

MacGregor told me with emotion that Greenpeace had used dirty tactics and told lies. He felt personally affronted by their attacks on the industry that he represented. He argued his case well. In his view, the Bering Sea pollock fishery has evolved into one of the cleanest and best-managed fisheries in the world, and he was proud to be part of it.

In the late 1990s Greenpeace came out against the industrialization of the world's fisheries and accused the new breed of super factory trawlers of "strip mining" the ocean with three-quarter mile wide nets.[4] This charge initiated a feud that continues to the present.

During the midst of the battle, Alaska's Republican Senator Ted Stevens, who already had a strong anti-offshore trawler viewpoint, and Greenpeace found themselves to be unlikely bedmates in the fight against the at-sea trawler processors. Like Greenpeace, Stevens was opposed to the factory trawlers partly because of pro-environment sentiments.[5] But his stance against the factory trawlers also reflected his pro-Alaska and anti-Seattle sentiment, and was one that supported his political backers. The factory trawlers were all based in Seattle. In December 1997 Stevens introduced a bill in Congress that effectively would have phased out factory trawlers. In answer, Bernt Bodal, CEO of American Seafoods, said, "I feel this is another tool for a fish grab."[6]

The 1997 descent from the bridge wasn't the first publicity stunt Greenpeace had performed. In August 1995 Greenpeace activists, in protest of overfishing and in promotion of a ban on factory trawlers, found a unique way of preventing the ships from leaving for the fishing grounds. Seven activists chained themselves to a trawler for ten hours and were arrested for criminal conspiracy. Six other protestors rappelled from ropes off the Aurora Bridge with a sign saying "Factory trawling guts family fishing."[7]

One morning in August of 1996, four divers entered the water of Puget Sound at 3 a.m. and wrapped chains around the propellers of five ships belonging to American Seafoods.[8] Later, four Greenpeace members dangled from the side of one of the ships on climbing ropes. Another group formed a floating human blockade across the channel exiting the pier to prevent the departure. Continuing their protests, in January 1998 Greenpeace members in rubber boats circled around the trawlers as they left for the Bering Sea in a symbolic display but did not attempt to stop them.[9]

Greenpeace activists also took their protests into other oceans where American Seafoods was operating. In 1997 the *American Monarch* had the highest capacity of any fishing boat in history. With a length of 311 feet, over a football field long, she could process about 1 million pounds of fish a day. Although the *American Monarch* was home ported in Seattle, she was never able to fish in American waters because the hull was built in Norway. Greenpeace raided the *Monarch* and hung a banner off her superstructure calling for the ban of factory trawlers. Fred Munson of Greenpeace said, "These ships are literally designed to overfish the world's seas, to deplete one fishery, and then move on to another ocean."[10] The *Monarch* was denied permission to fish off the coasts of Chile, Peru, and Argentina and sat idle for nearly a year.

Down in Chile, American Seafoods wanted to fish for southern blue whiting, which at the time was the object of a Chilean joint venture fishery with the Japanese conglomerate Nippon Suisan. The battle to block the *Monarch* went all the way to the Chilean Supreme Court, with American Seafoods emerging as the loser. It was rumored that some powerful Chilean politicians had competing financial interests. In 1998 the *Monarch* sailed for the western Bering Sea under a lease agreement as a bare-bones charter with a Russian company to fish for pollock there. Various reports indicate that fishing for pollock in the western Bering Sea was almost unrestricted and largely targeted on juvenile fish, or alternatively, that the Russians allowed an unsustainable 40%–50% of the pollock stock to be harvested there.[11] Since that time, the *Monarch's* name changed to the *Atlantic Navigator* and it now fishes, as part of Kjell Inge Røkke's company Aker Biomarine, for Antarctic krill in the southern oceans.

The pollock industry has counterclaimed that Greenpeace was telling lies and exaggerating. Each side has dug into its trench for the battle. For example, to this day, one of the major issues of contention is whether a 747 jetliner, or even how many of them, would fit into a trawl net. During my interviews I heard impassioned arguments about whether this was possible or not from former Greenpeace activists and from industry representatives, or whether the airliners had to be disassembled first. Obviously this was a tidbit that stuck in the craw of both parties. Technically, I think that since the wingspan of a 747-400 is about 60 meters, and a ship of only 1,500 horsepower could tow an Aleutian wingtrawl midwater net with opening dimensions of 73 x 64 meters, a 747 could fit inside. Commonly, large midwater trawls are over 100 meters wide. The so-called megatrawl was developed for the pollock fishery and has a 90-meter vertical mouth opening. The American Dynasty fished with a NET Systems Inc. trawl off bottom with a net that was 68 meters wide and 42 meters high.[12]

In 1998 Bernt Bodal of American Seafoods was reported to say, "Once again Greenpeace is making exaggerated and even false claims in an effort to alarm the public and enlist support for their cause. The reality is that a certain number of fish are going to be harvested—the number and size of the boats competing for those fish are irrelevant to the health of the stocks."[13] NMFS biologist Lowell Fritz said, "The fish don't care if they're being caught by a big boat or a little boat."[14] At the time, this was the conventional wisdom, but the number and size of boats is now a factor to consider in the management of the fishery because the larger boats can travel farther across the range of the species and can stay for longer trips than the shore-based boats.

Greenpeace aimed for a 50% reduction of the global fishing fleet and a complete ban on factory trawlers in US waters.[15] Young Fred Munson was one of the local leaders of the Greenpeace movement in Seattle. Munson attended West Chester University in Pennsylvania, majoring in economics and philosophy. Like many young Greenpeace activists, he had a liberal arts background, but a strong interest in the environment. When he started working on fisheries issues, he had to become fluent in the language of fisheries science to argue his point effectively. Munson started working at Greenpeace in 1986 and got involved in a number of their campaigns, such as recycling and prevention of ocean dumping. Then in 1993, he moved to Seattle to work on their fisheries campaign. When I interviewed Munson in 2011, he was the executive director of a private foundation with a conservation agenda. His office was a sanctuary of tranquility, with lights low and a nice view of the trees and water. Munson was calm and peaceful, and it was hard to envision the firebrand activist of his youth inside the man sitting before me. It had been almost twenty years since he worked with Greenpeace. We stumbled along through the beginning of the interview, but he began to rev up and recall more as we proceeded, and provided quite an insightful and balanced view.[16]

Munson recounted the days of conflict with the industry. The big issue was to initiate a ban on factory trawlers. Greenpeace was interested in this sector of the fishery because of the huge amount of capital invested in exploitation of the resource. The argument goes that once such a large amount of funds gets involved, then the issue of overfishing gets politically hot. Such a large investment by the industry drives them to keep harvesting fish to pay off loans. Jobs and economy become issues to vulcanize public opinion. Politicians become enamored with industries that make campaign donations, and the fisheries become "too big to fail."

Another issue that interested Greenpeace at the time was the individual transferable quota (ITQ) or catch share system. They understood the need to control the "race for fish," but it was developing too quickly and without a lot of dialog and consideration. In addition, the quotas were worth a lot of money and they felt the shares would likely be controlled by corporate interests in the long term. This would again promote a muscular industry and "politicalization" of the fisheries decision-making processes.

For the Greenpeace activists, working with the NPFMC was frustrating because the process was not set up to protect the environment. Many of the people involved in the council are associated with industry, and although they may give the appearance of being fair, there is still an economic interest involved. The council process seems to be heavily influenced in supporting the investment that has been made in the fisheries infrastructure.

The antitrawler campaign of Greenpeace was joined by Alaska Senator Ted Stevens, who started legislation to phase out the large factory trawlers and foreign ownership of them; it was clearly designed to hit American Seafoods hard. The antitrawler campaign was the first round in the salvo of shots fired by Greenpeace across the bow of the pollock fishing industry. The result was a big splash that drew a lot of attention. The next shot came a lot closer to the target, if not scoring a direct hit, in 1998.

## Sea Lions Bite Back

A death knell seemed to be sounded for MSY in marine fisheries in the 1980s as word of Peter Larkin's article, "An Epitaph for the Concept of Maximum Sustainable Yield," spread.[17] Larkin's poem poked fun:

> Here lies the concept, MSY
> It advocated yields too high,
> And didn't spell out how to slice the pie.
> We bury it with best of wishes
> Especially on behalf of fishes.

But the epitaph was premature. MSY rose from the ashes in 1996 as it got codified in the Sustainable Fisheries Act, which reauthorized the FCMA. The new legislation was now called the MSA, or Magnuson-Stevens Act.

Several staffers in the Greenpeace oceans campaign had worked in the Greenpeace anti-logging spotted owl program. From this experience they learned that you could "sustainably" harvest a forest and still destroy the natural ecosystem.[18] They, like the early fisheries scientists Hjort and Huntsman, recognized the parallels with logging practices. Their insight was that by considering only MSY to determine pollock harvest levels, NMFS was out of compliance with the National Environmental Policy Act, or NEPA. In particular, Steller sea lions (SSL) were being listed as an endangered species and nobody in the fisheries regulatory agency was doing anything about it. They recognized that the situation was a natural extension of what was learned in the forest logging campaign, that you could harvest a forest for MSY while endangering the habitat of the spotted owl.

In the 1950s the population of Steller sea lions numbered in excess of 240,000 animals, but by the late 1980s it had declined by almost 90%. The species was listed as threatened under the US Endangered Species Act (ESA) in 1990 and was listed as endangered in 1997. To date there has been no definitive cause attributed to the SSL decline, although several hypotheses have been put forward, including (1) nutritional stress (competition between SSL and fisheries for prey); (2) climate shifts; (3) predators, including killer whales; (4) contaminants; (5) disease; (6) incidental take by fishermen; and (7) directed hunting by Alaska Natives.[19]

Greenpeace first filed suit over the Steller sea lion and fishery conflict in 1992, citing the increased intensity of the trawl fishery and its impact on pollock as prey of SSL. There were several pieces of federal legislation that played an important role in the outcome of this suit and those that followed. The National Environmental Policy Act of 1970 required that any major federal action taken that would significantly affect the quality of the environment must be accompanied by a comprehensive examination of those potential effects, and must provide reasonable alternatives.

Another important piece of legislation was the US Endangered Species Act of 1973 (ESA), which established a procedure by which endangered and threatened species, and their ecosystems, may be conserved. The ESA mandated that decisions must be based on scientific data and criteria and not the effect of remedial actions on economic interests.[20] Finally, the Marine Mammal Protection Act (MMPA) of 1972 established rules by which the agency would develop procedures to govern interactions of fisheries and mammals, and prohibited killing, hunting, and harassment of marine mammals.

Prompted by the initial suit, NMFS implemented restrictions on when and where fishing occurred in the sea lion habitat. No-trawl regions of 10–20 nautical miles were established around sea lion rookeries, harvests were limited to quarterly quotas to spread effort over time, and separate quotas were established by region to spread effort across space.[21] But then in 1996 NMFS concluded in a Biological Opinion (BiOp) that the fishery management plan did not jeopardize Steller sea lions.

In 1997, studies were conducted to show separate eastern and western populations, and the western population was listed as endangered. Also in that year, the NPFMC established a total allowable catch of pollock that increased by 60% in the western portion of the Steller sea lion range. This increase in the harvest level prompted action by the environmental groups. NMFS was in charge of both managing fisheries and protecting SSL. In 1998, Greenpeace joined with the American Ocean Campaign, Earth Justice Legal Defense Fund, and the Sierra Club under a civil action accusing NMFS of failing to prevent jeopardy to SSL and failing to protect its habitat. There were violations of both NEPA and ESA. The plaintiff organizations contended that the NMFS managed the fishery to maximize the quantity of fish caught without enough regard for the protection of Steller sea lions that are using pollock as prey.[22]

The ensuing litigation over Steller sea lions and the fishery spanned six years and four trials. The ability to prove whether fisheries did or did not play a role in the threatened status of Steller sea lions was in dispute. The uncertainty of science was uncloaked. However, the court did not demand certainty, but caution, consistency, and a transparent logical thought process. The trials showed that the agency failed to analyze and rationally apply the data it possessed, and it failed to develop reasonable and prudent alternatives to avoid jeopardizing the critical habitat of SSL. According to MacBeath, the trials "applied the pioneering statutes of the American environmental movement to the complex issues of a complex fishery. By doing so, this case has established a precedent that likely will be used to evaluate future challenges to biodiversity in the North Pacific ecosystem."[23]

In 2000, the presiding judge of the litigation declared that Greenpeace had proven that the pollock harvest was "a reasonably certain threat of imminent harm" and closed critical habitat to the trawl fishery. NMFS responded with a reasonable and prudent plan to mitigate harm to the SSLs, and the judge allowed trawling to resume. But the plan included large-scale closures of critical habitat, which was unacceptable to the fishing community. The trawl industry sought the assistance of Alaska Senator Ted Stevens, chair of the powerful Senate Appropriations Committee. The industry prepared the text for Senator Stevens to limit the implementation of the new regulations, it was attached to an appropriations bill for Congress as a rider, and it eventually passed and was signed into law.[24] The rider mandated that the NPFMC, which is perceived by many to be controlled by the fishing industry, and NMFS consult together on measures to protect SSL. Some would say the industry control of the NPFMC is not just a perception; it is designed that way in the Magnuson-Stevens Act–Fishery Conservation and Management Act.

Steven's rider also included $20 million for research into the SSL problem and $30 million for disaster relief assistance to Alaska coastal communities affected by the fishery closures. Following passage of the rider, the NPFMC became a more equal partner to the NMFS in fisheries management and recovery of ESA-listed species.[25] The council formed a "reasonable and prudent alternatives" committee to propose changes to the management plan, which had eleven industry representatives and three environmental organization representatives. Greenpeace alleged that the committee was industry dominated.

The committee proceeded to weaken the mitigation regulations, and after a few more minor skirmishes, the courts allowed the pollock fishery to continue. The committee was allowed to come up with tools to craft regulations for each region, and a total package that would show no jeopardy to SSL. Since 2000, the SSL population overall in Alaska has increased slowly, but there is considerable regional variability. Throughout the western half of the western Aleutian Islands region, SSLs continue to decline, while in the eastern Aleutians and in the Gulf of Alaska, populations are stable or increasing. The pollock trawl fishery has been closed in the Aleutian Islands since 1999. In 2010 the western third of the Aleutians was closed to directed Atka mackerel and Pacific cod fishing, also important prey of SSL. Significant areas of critical habitat were also closed to both fisheries in the central third. This closure continues to be a contentious issue and in October 2011 was the subject of hearings in Congress. Representative Don Young of Alaska said, "The Fisheries Service is making regulations with limited information." He continued, "While we have no idea if these closures and restrictions will benefit the sea lion, we do know that they will have devastating effects on the fishermen and fishing communities. And out of fear of a lawsuit by extreme organizations, the agency hides behind 'the best available science' excuse and exercises an overabundance of precaution, akin to someone who can't swim refusing to bathe."[26]

In the long run, Greenpeace succeeded in forcing the agency to better protect the habitat requirements of the sea lions. In a broader sense, the lawsuits forced NMFS to give more consideration to ecosystem management. The legislative mandates of NEPA and ESA now had a serious bite, and the courtroom was another battlefield for the environment–industrial fisheries struggle.

## The Environmental Struggle Continues

Today, Greenpeace continues as the point of attack against federal policies in establishing fisheries quotas, and for protecting ecosystems. Generally, federal fisheries scientists and the pollock industry side together in defending harvest quotas based on single-species estimates of sustainable harvests, whereas Greenpeace and the activist group Oceana consistently insist on reduced quotas and broader ecosystem consideration. There remains a palpable animosity and lack of trust between the fishing industry and Greenpeace.

In an ongoing public relations war, Greenpeace asserts that the seafood industry has enlisted the National Fisheries Institute (NFI), which despite its federal- and scientific-sounding name is actually a trade association. In an "open letter to journalists," the vice president of NFI says "the news media has played into the hands of agenda-driven environmental activists and presented distorted reporting as fact."[27] Greenpeace says that the NFI's goal is to "kill the messenger" by attacking anyone who criticizes the seafood industry.[28]

Greenpeace continues to call for restraining the pollock fishery because of overfishing.[29] It urges a more precautionary approach to ecosystem management and the creation of marine reserves.[30] The At-Sea Processors Association counters that the pollock managers take a precautionary approach and set conservative harvest levels.[31] Furthermore, pollock fishing uses a midwater net that does not drag on the bottom, so the impact on the habitat is minimal, and bycatch has improved to less than 1% of untargeted species. There is almost no waste in the fishery due to a policy of full utilization.

While the fishing industry was fighting battles with the environmental movement at the gates of the citadel, there was a civil war going on within the industry that was equally strident and with much at stake.

· · · · ·

### SIDEBAR 4: LIFE ON A FACTORY TRAWLER

*Transpacific warned potential employees about the hard life on a factory trawler. If they were looking for adventure in Alaska, this wasn't the way. Their ship, the* Arctic Trawler, *had a crew of thirty-seven, and twenty worked in the factory. Each factory person worked two shifts a day, seven days a week. Shifts were four to six hours on, four hours off, when fishing was slow. If they were on fish, shifts were eight hours on, four hours off. When the boat went to fuel up, workers were given two hours of shore leave. Trips lasted three to four months. The company looked for workers with quick hands, mostly males in their twenties. Some workers were college graduates, and most had graduated from high school. Others were taking time off before returning to school. Several workers spent two years on board without going ashore except for two-week turn-around periods. The food was good. There were three entrees per meal and the cook was one of the highest-paid crew members.[32]*

*Arctic Alaska showed a movie to prospective employees. "Forget the romance of the sea," it said. "You'll be working in some of the worst weather conditions known to man. You'll eat, sleep, and breathe nothing but fish." It was reported in a lawsuit that a factory processor on the* Alaska Venture *worked shifts thirty hours long. Over one five-day period he was fed nine meals and allowed just fifteen hours of sleep. The ship's drinking water was contaminated with diesel and the vessel's air supply was filled with exhaust fumes.[33]*

*On board, the ship codend gets dumped and shoveled into a hatch where a conveyor belt runs it into the factory. The factory runs twenty-four hours a day, seven days a week. Machines chop off the head and slice out the guts. The valuable roe is separated out. The fish go through a line where a worker feeds them into a machine that either skins and fillets the fish or skins and* minces the fish into paste for surimi. The skin is the only part that isn't used. The guts and heads are pressed to extract oil, and then the rest, along with the skeletal remains, are made into fishmeal. Part of the captain's job is to keep the factory supplied with fish to maintain operations for twenty-four hours each day.

.　.　.　.　.

# 10 All in the Family: Olympic Fishing and Domestic Strife in the Industry

~~~~~~~~~~~~~~~~~~~~~~~~~~~~~~~~~~~~~~~~~~~~~~~~~~~~~~~~~~~~~~

*If he was going to have the last laugh, I was going to have it first.*
ARCHIE BUNKER in "All in the Family"

*I respect all fishermen. It takes guts to go to sea. And you have to be a little bit nuts to fish in Alaska waters.* ARTUR DACRUZ JR.[1]
~~~~~~~~~~~~~~~~~~~~~~~~~~~~~~~~~~~~~~~~~~~~~~~~~~~~~~~~~~~~~~

## Growing Pains

By the early 1990s the fishery had grown too fast for what was now recognized to be a limited resource. When the foreign fishery was being forced out of the US FCZ in the 1980s by the Americanization process, there was an opportunity to control the fishery, but by 1990 the fleet had grown to twice the size needed to catch the quota. The large number of vessels could catch the harvest quota in just seventy-two days. The fishing was on a first come, first served basis until the catch limit was reached.

The competition to catch the fish as quickly as possible is called "Olympic fishing" or "the race to fish." A respected fisheries economist at the University of Washington, Jim Crutchfield, said, "Several of us at the University of Washington pointed out that when we booted the foreign vessels out of the Gulf of Alaska and Bering Sea fisheries, that we had a wonderful opportunity, with only a few large harvesting vessels and mother ships in operation, to control entry so that we could keep capacity down to something approximating the yield capacity of the stock. We didn't do it. So we ended up with seventy-two large processing ships and a season that lasted only two or three months. Absolutely asinine!"[2]

In the late 1980s everyone was building ships at a breakneck pace. The factory trawler fleet increased from thirty in 1988 to forty-five in

1989 and fifty in 1990. Over twenty ships with US hulls were rebuilt in Norway. Røkke rebuilt the *American Empress, American Dynasty*, and *American Triumph*. Sjong, Uri, and partners converted another five ships. Erik Breivik rebuilt the *Pacific Glacier*.

Pereyra, along with several Norwegian-American trawler owner-investors, had the merchant vessel *President Wilson* converted into the 626-foot processor *Ocean Phoenix*. Part of the conversion was accomplished in Norway. The marine architects on the project were from the Norwegian company Fiskerstrand and Eldøy. In fact, beginning with the *Arctic Trawler*, Fiskerstrand and Eldøy did almost all of the architectural designs for the vessel conversions in Norway. The firm offered "package deals" which included vessel purchase, design, financing, co-investors, and shipyard contracts.[3]

In 1990 a group of thirty-five Danish and Faroese fishermen invested $25 million to buy Crystal Star Inc. and its two 230-foot freezer trawlers, the *Crystal Clipper* and *Crystal Viking*. These were originally built in the United States as offshore supply vessels, and were bought and taken over by Norwegian partners in 1988 and rebuilt in Norway as trawlers. The ships were later bought by Røkke and became the *Pacific Explorer* and *Pacific Scout*. It is said that one of the vessels had a Danish chef who made remarkable pastries and bread, and his contract was tied with the ship.

In the late 1980s, Oceantrawl, under Assen Nicolov (formerly of the *Golden Alaska*) as president and CEO, along with Japanese investors built the "bird boats" in Europe. The *Northern Eagle* and *Northern Hawk* were rebuilds completed in Norway (and another ship ended up sinking within sight of the dock in Norway[4]). The *Northern Eagle* was converted from a Hawaiian bulk sugar carrier to a 348-foot surimi processor in Norway. The *Northern Eagle's* sister ship, the *Northern Jaeger*, was converted in Germany.[5] Owing to the Reflagging Act, the clock was ticking on the deadline for when the bonanza of building ships in Europe would halt. Since the Norwegian shipyards were full, in order to beat the deadline, construction on the *Northern Jaeger* had to be done in Germany. The original US-built hull was from a "roll-on, roll-off" ocean transport ship, the *Inagua Ranger*, which carried trailer trucks between Florida and Puerto Rico. In 1996 Kjell Inge Røkke's companies took over management of these ships, and eventually their ownership.

The "sea ships" were built by the Norwegian owners of Emerald Seafoods after Reidar Sætmyr joined Lars Eldøy on a trip to the Gulf of Mexico in search of mudboats for conversion to trawlers. From 1988 to 1990 they built the *Claymore Sea*, *Heather Sea*, and *Saga Sea*. Seventy-five percent of the company was Norwegian owned and the rest was owned by Koreans.[6] Eventually they were bought by Røkke's company, American Seafoods, from the bankrupt Emerald Seafoods for $62 million. At that point they had been fishing in Russia for several years and never returned to fish in US waters.[7]

There were ships rebuilt in Asia and the United States as well. The *Traverse City Socony*, an oil tanker built in 1938, was converted to the *Bering Trader*, a 350-foot factory mother ship for salmon and herring in 1980. In 1988 it was converted to process pollock and groundfish for Kemp Pacific Fisheries and had from thirteen to twenty-three trawlers fishing for it.[8]

Francis Miller's company, Arctic Alaska Fisheries Corp., built their fleet mainly using shipyards in Louisiana and Alabama. Arctic Alaska started in 1981 with its first trawler, the 160-foot *NW Enterprise*. In 1985, Bender Shipyards of Mobile, Alabama, built five vessels for Miller to work groundfish off Alaska. These were the *Pacific Enterprise*, *Ocean Enterprise*, *Arctic I*, and *Arctic II*, as well as the *Arctic Enterprise*, which was a processor.[9] By 1992 Miller's company had thirty vessels. Individual boats were owned by investors, including many doctors and attorneys. Arctic Alaska offered a management contract through limited partnerships. Then it went public. In 1992 Arctic Alaska expanded into Asia, first to have fish processed, and then to both buy and fish there. They purchased fish in Russia, shipped them to China for processing, and then shipped them to the United States.[10]

Shore-based processing facilities were also being built at a rapid pace. Trident Seafoods built its first processing plant on the volcanic Aleutian Island of Akutan in 1981, about thirty miles from Dutch Harbor, and started processing pollock in 1983.[11] Then in 1984, the Japanese trading company Marubeni opened a small surimi line in Kodiak at Alaska Pacific Seafoods. In 1985 Universal Seafoods (Unisea) announced it would build a surimi plant in Dutch Harbor with Nippon Suisan.[12] This was followed by the announcement by Ward's Cove Packing Co. that they would partner with Taiyo to build a surimi plant in Dutch Harbor. In May 1985, American and Japanese interests set up a factory to produce artificial crab from pollock in Seattle. The company, Trans-Pacific Products, was a venture of Stewart Investment and Western Alaska Fish Co., a subsidiary of Taiyo.[13]

In 1986 Nippon Suisan Kaisha (also known as NISSUI) opened the Great Land Seafoods plant in Dutch Harbor.[14] That same year, Taiyo partnered with Marubeni and Columbia Wards to open Alyeska. Between 1988 and 1993, six onshore surimi processing plants were built in Alaska. Many of the plants were owned, partnered, or operated by Japanese companies.

### Onshore-Offshore Battles

Two distinct sectors began to form in the pollock fishery. The large Seattle-based ships that caught fish and processed them in onboard factories, or sometimes processed catches from smaller catcher boats, formed one group. The other group was the shoreside processors and the smaller catcher vessels that were too small to have onboard factories and the freezer capacity to stay at sea for long. Many of the smaller catcher vessels came from the joint venture fishery as the foreign processors were forced out during the Americanization. They now contracted with either the floating factory mother ships or the shoreside processing plants to deliver their fish. They were characterized as Alaska-based, but 80% actually were home ported in Seattle.[15]

Owing to the much greater fishing capacity of the large factory trawlers, they had an advantage over the small draggers. The catcher-processors had larger crews that could work 24/7, larger nets made possible by much greater towing power, greater range and storage, and the ability to work in rougher weather than the smaller boats. As their numbers increased, they had the capacity to catch the whole quota of pollock in about sixty days and were able to outcompete the small trawlers that supplied the shoreside plants.

The smaller boats, which had developed the skill and knowledge for catching pollock during the joint venture years, felt that they were getting squeezed out by the large factory trawlers, which were dominated by foreign investors. The factory trawlers had the northern part of the Bering Sea mainly to themselves. It was inefficient for the smaller vessels to fish in the northern Bering Sea because of their limited capacity and the long trip back to the Aleutian processing plants. But in the southern Bering Sea, the small catcher boats and factory trawlers competed for fish. This competition for a resource that was finite set off a series of skirmishes between the inshore and offshore harvesting sectors as the slices of pie got thinner and thinner.

The first major battle between the onshore and offshore components was in 1989 when several large factory trawlers caught most of the entire Gulf of Alaska pollock quota almost before the inshore catcher boats could get away from the docks. The factory trawlers swooped in and caught 40,000 tons of pollock in just eleven days. The Kodiak catcher boats supplying the shoreside processors caught only 20,000 tons of their expected haul of 60,000 tons, and 20% of Kodiak's workforce was put out of work for four months.[16] A representative of the factory trawlers said they did it because they heard that the Kodiak shore plants were "roe stripping"; that is, they harvested the valuable pollock eggs but discarded the rest of the fish.[17] They reported to a silenced chamber at a NPFMC meeting that they would do it again if they heard of more roe stripping, to which one council member responded to the speaker, "You have the balls of a gunslinger and the brains of a jackass." The truth is, in those days everyone was roe stripping.[18] The roe was the most valuable product and there was little processing needed. The idea was to get the most value with the least added cost and to capture as much as the quota as you could quickly. There was tremendous waste. One skipper reported that he'd converted his crab boat to a pollock dragger, and was so disgusted by the waste that he switched back to crabbing.

After the Kodiak incident, the local fishermen requested that the NPFMC consider separate quotas for the inshore and offshore sectors. This idea was opposed by the factory trawlers. But by now the factory trawlers were catching 80% of the northeastern Pacific pollock harvest, up from 65% the year before. A representative of the Pacific Seafood Processors Association said, "By the end of 1990, they [factory trawlers] will have captured 90% of the available pollock."[19] With regard to establishing separate quotas, Wally Pereyra said, "I think that's absolutely improper, and disregards the marketplace, the investment we've made, and the risk we've taken."[20] The head of the North Pacific

Vessel Owners Association also opposed it, saying, "We are strong supporters of the least amount of regulation, just a little to the right of
Genghis Khan."[21] A contentious debate ensued over the next two years
as the NPFMC prepared to act on the proposal.

That year the *Seattle Times* published an article reporting that public affairs consultants and Washington, DC lobbying firms had been
hired by both sides to present their viewpoints and to enlist support.
A battle ensued in the press to win public opinion. The American Factory Trawler Association ran an advertisement that read, "Japanese
processors [the shoreside processors] are forcing the United States to
give away what American fishermen worked so hard to gain." Dave
Fraser of the *Muir Milach*, a small trawler without refrigeration that
sold its catch to the factory trawlers, complained that we were giving the fishery back to the Japanese after the effort to Americanize
it.[22] The factory trawlers wanted to maintain the status quo and to see
who could catch the quota first in free competition. "The marketplace
works just fine," said a factory trawler representative.[23]

Bruce Buls of the American Factory Trawlers Association said about
the reallocation, "There's a billion dollars of boats and gear floating
around out there, thousands of jobs, and it's all in jeopardy. We have
no choice but to fight, and fight we will. . . . Our success has in a way
been our downfall. . . . We've been so successful that now everyone
else wants a piece of the action. They don't seem to be able to secure
enough through competition. They want it provided to them by quota.
. . . Part of the tactics of our opponents has been to try and characterize factory trawlers as rapers and plunderers. It's an image that the
public can also embrace, but I think it does a disservice not only to the
factory trawlers but also to the entire fishing industry."[24]

The Pacific Seafood Processors Association said that their smaller
boats could not follow the fish like the big ships: "we don't have the
horsepower." Besides, the shoreside processing plants needed a steady
supply of fish to maintain processing efficiency. Their response stated,
"Beware of the fish stories. The factory trawl fleet is trying to feed you
a very big line."[25]

John Iani of the Pacific Seafood Processors Association said the
large factory trawlers could "wipe out" the onshore industry if they
were not restrained. Of the allocation plan, he said that the factory

trawler owners were now "really squealing." He pointed out that the factory trawlers were discarding about 6% of the catch versus 1% for the catcher vessels.[26]

Iani said that the factory trawlers were getting only 12%–14% yield, while the shoreside plants were getting 20%–21%. The reason is that the factory trawlers had no incentive to maximize yield. They attempted to get as much value out of the fish as quickly as possible. There was a race to catch as much of the annual quota as possible and to extract the maximum profit. The shoreside plants tried to get as much value as possible out of each fish because they had to pay the catcher vessels for them.

In an unusual turnaround of philosophy, Alaska's conservative Republican politicians lined up favoring government regulation over the free market. Don Young, the Republican Congressman from Alaska, quipped, "We didn't create the 200-mile zone to Seattleize it. You are killing a community with an Americanized invader. You might as well bring the foreigners back."[27] Young neglected to mention that the foreigners were still there; 80% of the shoreside processing capacity was controlled by Japanese interests.[28]

Wally Hickel, formerly a cabinet official in the free-market Reagan White House, was the newly elected governor of Alaska. One observer remarked, "The new governor perceives Alaska as 580,000 square miles that need paving."[29] Hickel arrived in Alaska from a dust-bowl tenant farm, and was a Golden Gloves boxing champion with thirty-seven cents in his pocket. He became a millionaire who presided over opening the North Slope to oil drilling. Referring to the offshore fishing fleet, he said, "It's our ocean and our fish . . . and the trawlers they come up here and they do the same thing [overfish] in the Bering Sea. I'll get the environmentalists with me to help manage that . . . because they [factory ships] are raping the fishery. It's our ocean. It's our resource. If we don't do something with the trawlers, they'll destroy themselves economically. They'll wipe out and go."[30]

In 1991, NPFMC member Wally Pereyra brought up the topic of rationalization of the fishery at one of their meetings. As a graduate student at the University of Washington, Pereyra had taken a class from Jim Crutchfield in natural resources economics and learned about the "tragedy of the commons." Somehow, entry into the fisheries had to be limited or the fleet would be overcapitalized and the pressure on the resource would be intense. He proposed, "Maybe the way to do it is a public corporation that owns and manages the resource, and leases the rights, using the proceeds to pay the costs of management." He admitted though, "It's fear of the unknown, nobody is going to propose auctioning off fish. You'd be defeated before you started."[31] Pollock boat skipper Dave Fraser supported the idea when he said, "This is a public resource, like trees or minerals, why don't you sell it."[32]

Fraser said of the allocation politicking, "The process has turned into a horror show. It comes down to power, because there are no principles guiding their allocation decisions."[33] Bart Eaton, an executive at Trident Seafoods, said, "With the foreign fishery we had it easy. We could just say yes or no, and that was that. Now the council is frozen. This is a democracy, but sometimes it's like gargling peanut butter."[34]

In 1992 the Alaska delegation on the NPFMC had a 6–5 majority on the council. They voted to reserve a specific portion of the quota for catcher vessels and shoreside processors. They also voted to set aside 7.5% off the top of the reserve of pollock for a community development quota (CDQ).[35] The idea was to provide some economic benefit to the impoverished communities of coastal western Alaska. The idea for a community-based quota had been seeded in 1989 when the American Factory Trawlers Association (AFTA) set up a fund for villages, and formally announced the Bering Sea Commercial Fisheries Development Foundation in June 1991. This was a nonprofit to identify and finance projects designed for western Alaska coastal communities. The foundation was funded by an assessment on each ton of fish. They said the tax should generate $1 million each year. It was a public relations vehicle to show their involvement in the Alaskan economy. A spokesman of the AFTA said, "In addition to the $120 million our members spend in Alaska each year, we're pleased to add another dimension to our contribution to Alaska's economy."[36]

The formal CDQ concept was pushed along by Harold Sparck, a rural fisheries activist who headed a group called Nunam Kitlutsisti, which means "protectors of the land" in Yup'ik. He was joined by Henry Mitchell, a member of the NPFMC and fisheries lobbyist.[37] Mitchell said that the Seattle contingent opposed it (presumably because the AFTA already had their own foundation), but voted for it because they thought inclusion of the issue would cause Secretary of Commerce Barbara Franklin to reject the whole measure. Franklin ended up signing off on it under pressure from Senator Stevens. The CDQs are shared among six community coalitions representing sixty-five villages and 27,000 people.

From the remainder of the quota for the eastern Bering Sea pollock, 35% was reserved for the inshore catcher vessels delivering to the shoreside processors and 65% to the offshore processing vessels. The inshore portion was to increase to 40% the second year and to 45% the next year. An inshore operation area for the catcher vessels (CVOA) was also established in which the factory trawlers and draggers delivering to mother ships were not allowed to fish. Furthermore, 100% of the Gulf of Alaska pollock were reserved for the inshore catcher vessels. However, the NOAA administrator disallowed the increased inshore catches in years two and three.

Now the factory trawlers were hit high and low. Not only were there too many of them, but also their share of the quota was now decreased by one-third. On top of that, pollock stocks were at their lowest level in a decade, and the catch quota was sliding downward. Bob Morgan, president of the American Factory Trawlers Association, complained, "Most of the investment in the factory trawler fleet was encouraged by the government to a tune of $1.1 billion." Bart Eaton of Trident Seafoods added, "The problem is that the investment environment, encouragement, and so on have created downstream problems that were not understood."[38]

The new allocation of quotas set off a political row. As mentioned, the US Government Accountability Office (GAO) report had found that the majority of the shoreside processing plants in Alaska were owned by Japanese interests. In fact, of the current major shoreside processors in Alaska, Trident is the only American-owned company. (Icicle Seafoods has floating processors and a surimi plant in Washington.) Westwind and Peter Pan are partly or wholly owned by Maruha of Japan, and Unisea is part of Nippon Suisan. The GAO found that the factory trawlers were about 30% foreign owned and they employed a higher percentage of Americans to work the lines than the shoreside processors.[39] Representative Jolene Unsoeld from Washington said, "It looks like blatant politics," and blasted the decision as a power play favoring Alaska.[40] Lawsuits and countersuits were filed. One company owning factory trawlers stated that restrictions were "an illegal scheme to deny Americans their fundamental rights in order to further foreign investments in Alaska and to provide an unfair competitive advantage to foreign-operated businesses there."[41]

In another development of this period, Tyson Foods, the world's largest chicken growing enterprise, bought Arctic Alaska Fisheries Corp. in 1992 for $212 million.[42] Arctic Alaska President and CEO Ron Jensen said of the merger, "It's a marriage made in heaven." Arctic Alaska wanted to take advantage of Tyson's distribution system and retail market. Francis Miller was to continue as chairman of the board.[43]

Don Tyson was known as a shrewd and folksy character, and a risk-taking bare-knuckled businessman who skirted the edge of the law.[44] "I'm just a chicken farmer," he liked to say.[45] He had a penchant for khaki uniforms, and aimed to put Tyson "at the center of the dinner plate." Don Tyson was to learn that catching and processing fish are not as predictable as growing fryer chicken. Tyson was good friends with President Clinton, raising speculation that one processing sector or the other would gain political favor. Tyson's assets acquired from Arctic Alaska included thirty-one fishing vessels, one floating processor, and also shoreside plants in Alaska, Seattle, and Oregon. But Tyson also now owned five offshore catcher processors as well. Tyson further inherited an expensive lawsuit from the sinking of Arctic Alaska's the *Aleutian Enterprise* in 1989 with the loss of nineteen lives; see Sidebar 2), and fines from the US Environmental Protection Agency and the State of Alaska for illegal dumping.[46]

Tyson was having a hard time turning a profit in the fishery business with the erratic supply of fish and worsening market conditions. This prompted him to say euphemistically, if not poetically, "It's like making love, sometimes . . . it happens sooner rather than later, but sooner or later, it's going to happen."[47] It never did happen for Tyson. Fluctuating supplies of pollock, inherited liabilities from Arctic Alaska, plummeting prices for whitefish, and increasing regulations wilted Tyson's own expectations for a happy climax. In 1999, Trident Seafoods bought out Tyson's ships and shore holdings.

By 1993 the situation of overcapacity in the trawl fleet was critical. Both the Alaska- and Seattle-based fleets had overgrown the resource. The number of days to catch the entire offshore quota decreased from 148 to 29 days between 1991 and 1994, and the inshore fleet filled their quota in 41 days in 1994 as opposed to 148 days in 1991.[48] In 1993, low prices, short seasons, and small-sized fish, which are less valuable, plagued the fishery. Because of their debt burden, the big ships needed to keep fishing. But many vessels remained tied up in port for months, while others headed to the Russian side of the Bering Sea to fish for pollock.[49]

Meanwhile, declines in the price of pollock in Japan in 1993–94 resulted in declining profits. Nine ships in the offshore factory trawler fleet of sixty-four fell into bankruptcy.[50] By 1998, twenty factory trawlers had gone bankrupt.[51] Kjell Inge Røkke was able to buy twelve of these. Røkke was speculating that an individual quota share system would be implemented and the catch history of the ships was important, so he was essentially buying quota of pollock for the future.[52]

Christiana Bank of Norway was known as the "go-to bank of Alaskan fisheries." It was Norway's second largest bank, and was taken over by the Norwegian government in 1991. By 1994, 80% of their fishing loans in the North Pacific were in technical default, and four factory trawlers had already been foreclosed. The bank now demanded that vessel owners pledge the fishing rights that might one day be granted to the vessels as collateral if they wanted refinancing. This was a harbinger of what was to come. The rights to gain access to the harvest quota would be worth hundreds of millions of dollars, and Christiana wanted a stake in them, which they apparently gained through close association with Røkke.[53]

Since they opposed a quota system dividing the catch between the onshore and offshore sectors, the factory trawler fleet complained that the 1992 agreement, which set aside 200,000 tons of pollock for Japanese-owned or controlled onshore processors, was at the expense of the factory trawlers. They reported that 2,000 jobs in the Seattle region were lost as a result of the onshore/offshore allocation. By now Washington State's powerful Senators Henry Jackson and Warren Magnuson were gone, leaving relatively junior members Slade Gorton and Patty Murray to take up the slack. Up north, Alaska had Senator Ted Stevens and Congressman Don Young, who were vociferous supporters of the shoreside processors. The trawler community complained, "What we don't have is a level playing field."[54]

The inshore/offshore quota agreement was set to expire in 1995 when a "free for all" would ensue. The council passed an amendment renewing the quota allocation, but it reduced the inshore operation area for the catcher vessels (CVOA), and allowed catcher processors in the area if the inshore catcher vessels had already filled their quota there. By this time the council had adopted a limited license program, and was considering some form of individual fishing quota (IFQ) plan.

An IFQ guarantees a fisherman of having an exclusive right to a share of that year's harvest quota. There is a cornucopia of variations in IFQ contingencies. But in general the "catch share" is attached to the fisherman or the "steel" (the vessel), or sometimes even the company, or sector (defined by type of gear, size of vessel, where processing occurs, or even by some other parameter). In general, the catch share is transferable because it can be rented, leased, or sold.

In 1996, under heavy lobbying pressure from the environmental coalition, Congress passed the Sustainable Fisheries Act (SFA). This bill provided protection from habitat destruction and implemented plans for ending overfishing, for annual reporting of overfished stocks, and for rebuilding them; and it required that the fisheries yield must be aligned with MSY rather than with short-term economic or social gains. Codifying the concept of MSY was important and will be discussed again later. Reducing bycatch and waste were also addressed.

According to Michael Weber, the Sustainable Fisheries Act was an important shift in the policies of fisheries. Previously, the fishing industry, backed by the federal government, was unchallenged in setting national policy in fisheries, with the aim of expanding fishing rather than stewardship of the resources. In the passage of the SFA, environmental groups flexed their newly found muscles to shift emphasis from development to conservation.[55]

In 1997 the NPFMC made more changes to the inshore/offshore quota division. Each time changes in the quotas were proposed, major fireworks were set off among the interested parties because millions of dollars, even hundreds of millions, were carried in the sway. There was a great diversity of opinions about the IFQ plan depending on whether the sectors perceived they were going to make or lose money. In 1997 the council increased the inshore sector's quota to 39%.

As more dramatic legislation loomed ahead, the lines separating the different interests were clearly drawn. In 1997, Dave Benson of Tyson

Seafoods said, "The stakes in this are very high. Companies affected by this legislation are going to spends lots of money [lobbying] as this bill moves through Congress."[56]

## The American Fisheries Act

During the discussions of the inshore/offshore allocations in 1997, Congress was considering legislation to stiffen rules about foreign ownership. This was a pointed attack on American Seafoods, which was owned by the Norwegian Kjell Inge Røkke and other Norwegian investors. The American Fisheries Act was introduced in the Senate by Ted Stevens in 1998. The prime supporters of the act were Tyson Seafood Group and Trident Seafoods, which would have greatly benefited if passed as originally written.[57] This legislation would have stripped rights to fish in American waters from numerous factory trawlers built in foreign countries (mainly Norway) under the Anti-Reflagging Act exemptions. It would further limit fishing vessels to 165 feet and limit foreign ownership in US vessels to 25%.

Stevens was particularly incensed about fishermen avoiding the requirement of having American-built ships by taking old US-built hulls and having them completely rebuilt in Norway. In particular, referring to one of American Seafoods' ships, he said in testifying before a Congressional Committee, "Now take a look at this next one [referring to a photograph] . . . the *Acona* after being rebuilt. A 74-foot 167-ton vessel is now 252 feet, weighs over 5,000 tons . . . and it's now called the *American Triumph*, it was the *Acona*. The only thing left of the *Acona* is a piece of steel in the side. I'm told there may be two pieces of steel, one in each side. Now that's a rebuild. It really is a totally new Norwegian vessel brought in and now poses as an American vessel. And it is flagged as an American vessel. This is one of about twenty of the so-called rebuilt vessels that now fish in the Bering Sea off our State."[58] Stevens went on, "We're talking about stealth foreign vessels in our waters flying the US flag. . . . I think this is one of the scandals of the fishing industry, what happened during this period . . . and incidentally, I'm not going to rest until I get to the bottom of that scandal. I believe there was really fraud." As it turned out, the *American Triumph* was one of the lucky ships granted a quota to fish indefinitely.

In 1997, the acrimony between the shore-based processors and catcher processors was so great that Chuck Bundrant, president of Trident Seafoods, and Peter Stitzel, vice president of American Seafoods, got into a wrestling match in the parking lot behind the Doubletree Hotel in Seattle while attending an NPFMC meeting. According to an article in the *Seattle Post-Intelligencer*, Stitzel was holding a sign with Bundrant's picture and images of two Japanese company CEOs on it. They scuffled over control of the sign. The grappling seafood industry leaders were told to leave the premises by Doubletree Inn's security.[59] Stitzel commented afterwards, "It was childish on both of our parts." But Bundrant wouldn't comment. Afterwards, the hotel security staff insisted that a guard be present at the meetings.[60] Interestingly, it's said that Bundrant moved to Alaska after watching the hard-fighting movie star John Wayne in the film "North to Alaska" while he was a college student in Tennessee.[61]

Chuck Bundrant is an example of one of the most successful American fishermen. Bundrant has been described as a titan of the fishing industry.[62] In 2008 his company hit over $1 billion in sales. Bundrant grew up in Tennessee and planned to be a veterinarian. In 1961 he set out for Alaska and got a job on a processing ship in the Aleutians. He worked his way up to deckhand and then captain, and bought his own crab fishing boat in 1965. Bundrant has been described as a man of "unwavering convictions" of how things should be. He "inspires absolute faith from friends and sometimes fear among those who oppose his interests."[63] Once, two men in their twenties who were working on a boat skippered by Bundrant complained about the conditions of working for him seventeen hours a day. Bundrant handed them survival suits and told them to jump into twenty-foot seas to be picked up by another boat and taken back to shore. Bundrant said this was the normal procedure and that he had done it many times.[64] On the business side, Bundrant had a great vision of the future of the industry and took huge risks to enact that vision.

Prior to 1998 it looked like Senator Stevens and his friend and supporter Bundrant were opposed to IFQs, as they were afraid the Seattle trawl fleet would gain an unfair advantage.[65] Bundrant had a particularly close relationship with Stevens. One report reads, "Bundrant's good fortune has not come from the magic of the free market alone. Over the years, Bundrant cultivated some strong allies in Congress—most especially, just-deposed Alaska Senator Ted Stevens and representative Don Young."[66] In the late 1970s and 1980s, one of the Trident boats employed Ted Stevens' son Ben as a crewmember, and in the mid-1990s Trident employed him as a Capitol Hill lobbyist.[67] There was a complicated web. Stevens' main staffer on the American Fisheries Act (AFA) legislation, Trevor McCabe, became Ben Stevens' business partner and co-owned land with Congressman Don Young's son-in-law. Trevor McCabe is now a well-paid executive of Coastal Villages, the largest of the community development quota (CDQ) organizations. Bundrant and Trident knew how to play the political game. Bart Eaton, a partner in Trident, was reported to have a long-standing friendship with Stevens; and Trident employed Brad Gilman, a former Stevens aide, as a lobbyist.[68] The chief counsel for Trident, Joe Plesha, was a former staffer of Senator Murkowski and had worked with Stevens on various pieces of legislation.

The Congressional staffers working with industry lobbyists played a major role in the legislation. One former lobbyist told of a discussion with a coworker after a prominent member of Congress made a statement on the legislation that seemed off-base. He said, "He's an idiot . . . but he's our idiot."

Bundrant's opposition to the IFQs stemmed from not being able to see how it would benefit Trident substantially. In fact, judging from the experience of another company, Icicle Seafoods, in the halibut and sablefish IFQs, the quota share system had the potential to hurt Trident. Bundrant appeared to want some quota to be attached to the shoreside processing facilities, rather than solely to ships and individuals.[69]

In the autumn of 1998, Senators Stevens and Gorton invited pollock industry representatives to Washington, DC, to consider the allocation issue, foreign ownership, and foreign-built trawlers in parallel. Who actually attended the meeting is somewhat controversial, with some people reporting that everyone with a major stake was there, and others saying there was exclusion, especially of most of the NPFMC, which was supposed to be regulating the fishery. Stevens' main focus was on ending foreign ownership and the reign of the large catcher processors, but since IFQs also carried a lot of economic weight, that seems to have been a major issue. Bundrant's desire was for quotas to be tied to the processors, but when a structure for co-ops was proposed such that each processor would run a cooperative of catcher vessels that was tied to quota, that seemed like a good alternative.

In the end, the negotiations produced results that had something for everyone at the table. The American Fisheries Act (AFA) was attached as a rider to a general appropriations bill for the entire US government in 1999. Stevens was known to use such riders to avoid the legislative process and associated political debate; because the riders are attached to other more important legislation, they pass without much discussion.[70] Once passed, the riders become law. Passing the AFA in this way restructured the pollock fishery without involving the NPFMC, NMFS, nongovernmental organizations (NGOs), many fishermen, or fishing communities in the discussion.[71] It would take "an act of Congress" to overturn this legislation.

Greenpeace had made passage of the AFA a cornerstone of its global campaign to eliminate factory trawlers.[72] But after first supporting the AFA, they and several other environmental groups came out against the legislation, contending that it was constructed in a cozy and nonpublic way, and did nothing to protect the Bering Sea ecosystem. Moreover, the new allocation would shift catches to the onshore component, putting more pressure on the habitat of the endangered Steller sea lion.[73]

The long-term goal of the AFA was to phase out factory trawlers larger than 165 feet, and especially the foreign rebuilds. The prime sponsors of the bill were those companies that would directly benefit from its passage. These benefactors included the industry giants Tyson Seafood Group and Trident Seafoods. The big winner of the future legislation looked to be Tyson Seafood Group[74] (later to become a part of Trident Seafoods) and the shoreside processors. If passed as originally written, it would have greatly weakened their primary competitor, American Seafoods, and would have put some other smaller factory trawlers out of business. Trident was now Alaska's largest shore-based processor, while Tyson had a fleet of eight US-built factory trawlers as well as catcher vessels. The two companies, although united on the AFA, were bitter rivals over onshore/offshore allocations.[75]

The primary intent of the AFA (SB 1221) was to correct loopholes in how the US Coast Guard enforced the 1987 Commercial Fishing Industry Vessel Anti-Reflagging Act. The original Anti-Reflagging Act had two purposes: that US citizens should have a controlling interest in the catcher-processor vessels, and the vessels should be built in US shipyards. That is, no more "foreign face-lifts." But there were certain conditions whereby owners were allowed to operate vessels that didn't meet the requirements of the act. And vessels that had begun to be rebuilt or had contracts in place, if finished by a certain timetable, would receive the fisheries endorsements.[76] These were bothersome loopholes in the legislation.

The object of Stevens' earlier wrath, the *American Triumph*, was later found to have a letter of sale on the vessel filed by the owners that was illegally backdated to meet the Anti-Reflagging Act timetable, and the Coast Guard was about to strip the *Triumph*'s fishing endorsement, but ironically, wording incorporated in the AFA specifically stipulated that the ship could legally fish in the Bering Sea, trumping the Coast Guard's action.[77]

The big winner of the AFA legislation ended up being Trident Seafoods. The Akutan Catcher Vessel Association, which sold its catch to Trident, landed with more than 30% of the shoreside allocation, more than any other co-operative group. That year Trident spent $150,000 lobbying Congress in its behalf.[78] But the offshore sector also considered the AFA as a victory, because now they had a guaranteed quota, and one that was split off from the mothership fleet.

There were others who were kept away from the table who were bitterly disappointed. One fisherman said, "The ones who overcapitalized the most, who showed very often the least business sense, are the ones who stand to gain the most. No matter how they treated this resource, regardless of their attitude toward this publicly owned thing, we're going to give it [the quota for pollock] to them forever."[79]

The AFA had several critically important aspects that influenced the pollock fishery, all involving access to the fishery. First, it put new rules on ownership, furthering the process of Americanization. Foreign-owned factory trawlers were forced out of the fishery. On the other hand, foreign-owned shoreside processors were allowed to remain. Second, set quotas were allocated to the three sectors of the fishery. Ten percent was reserved for the CDQs off the top. Of the remaining quota, 50 percent went to the inshore catcher vessels, 40 percent to the offshore factory trawler processors, and 10 percent to the mothership catcher vessels. Finally, the fishery was closed to new entries.

Within the details of the AFA, there were also important components that have greatly influenced the fishery. First was the establishment of fishery cooperatives. These are arrangements by which the sectors could organize and divide up the sector's share of the quota. Also, access to the quota could be leased, so not all boats had to be fishing. This increased efficiency.

The "seven magnificent trawlers," also known as the Gang of Seven, were independent catcher vessels that had been delivering pollock to the factory trawlers. They were given a quota of 3.4% of the total allowable catch (TAC) of pollock. All but one of the Gang of Seven now lease or exchange their quota to the factory trawlers through a cooperative agreement, thus profiting from the AFA without ever dipping a net in the water. That's worth about $10 million to the vessels. Fishermen grumbled. Dave Fraser of the *Muir Millach* defended the leases as a just reward for "homesteading" the Bering Sea.[80] Pete Stitzel of American Seafoods said, "The richest co-op fishermen might not fish at all, he will lease or sell his boat and associated quota share. A boat can sit on the beach and make $250 per ton of pollock, and the owner has no cost, none but the moorage. A lot of fishermen I know are trying to sell out right now."[81]

But not everyone was happy with the new arrangement. "Ordinary fishermen won't be able to buy into a co-op. Only big processors and fishing corporations willing to pencil out investments over many years or even decades will buy boats and quotas now soaring in value."[82]

As a provision of the AFA, nine offshore catcher processors, which were responsible for about 10% of the TAC, were "bought back" by the inshore processors. They were provided with a $75 million federal loan to do so and a fee was implemented to repay the loan. Eight of the nine bought-out vessels were scrapped to prevent them from entering other global fisheries. The other twenty catcher processors allowed to continue fishing pollock were also restricted from spilling into other US fisheries by so-called sideboard amendments.

Other big winners from passage of the AFA were the CDQ associations. Their quota was increased from 7.5% to 10%, now worth about $30 million per year.

The "Race for Fish" was now over. The Pollock Conservation Cooperative (PCC) formed the offshore cooperative group. The companies in the co-op were American Seafoods Company, Glacier Fish Company, Tyson Seafoods Group (now Trident), Alaska Ocean Seafood, Alaska Trawl Fisheries, Arctic Fjord, Arctic Storm, Highland Light Seafoods, and Starbound Partnership (Aleutian Spray). In the mothership cooperative, there were now three mother ships. The nearshore sector included 110 catcher vessels delivering fish to six shoreside processing companies.

The catch share program had a sunset date of expiring in five years, or in 2004, but once again without a Congressional hearing Stevens added a rider to a spending bill to eliminate the sunset date, vesting the rights to pollock indefinitely.[83] Industry officials praised the measure, saying it would show lenders that the business is stable, making it easier to get financing. Alaska conservationists said there should be some return to the public in exchange for exclusive rights to a public resource.

The formation of cooperatives has benefited both the industry and the resource. Under the agreement of the PCC, fishermen limit their catch to a specific percentage of the TAC that is allotted to their sector, the offshore trawlers. The pace of fishing is more deliberate, allowing the processors to achieve a higher quality of product, thus attaining a better price and a higher yield from the fish. Catches are reported to the NMFS daily, and two federal fishery observers are onboard at all times. The co-op pays for the cost of the observers, who are contracted through private companies.

Catcher vessels delivering to shoreside processing plants have an area nearshore that is banned from the factory trawlers. The catcher vessel quotas are each linked to a cooperative centered around a processing plant. Since the catcher vessels are small trawlers, there is only one observer onboard each vessel.

Management of the pollock fishery is now highly structured and intense. NMFS and the NPFMC set season opening and closing dates, restrict catches to specific areas and times, and set bycatch limits. Daily bycatches are reported to a private company called SeaState, which tracks the bycatch of each vessel and sets rolling area closures in "hotspots" where the bycatch is found to be high. In general, this process has been effective. Bycatch rates are now less than 1%.

• • • • •

SIDEBAR 5: DUTCH HARBOR–UNALASKA, ALASKA

*The community popularly known as Dutch Harbor lies midway out in the Aleutian Islands chain. It is actually two small communities that have grown together as Dutch Harbor-Unalaska. The history of this place is* closely linked to the pollock fishery, both physically and symbolically. Dutch Harbor is the largest fishing port in the United States on the basis of tonnage landed. Both the fishery and Dutch Harbor endured some turbulent years and now both have calmed down. Unalaska has a habit of getting swallowed up by events that bring outsiders to it. There was a Unangan village on the site of the town for several thousand years. Then it became an important outpost in the Russian fur trade. With the arrival of Europeans, most of the native population was decimated by epidemics. Other natives were kidnapped and then enslaved to work on the Pribilof Islands harvesting fur seals. World War II saw an influx of 10,000 soldiers and a forced evacuation of the native people to camps in southeast Alaska. Large numbers died under the poor conditions of confinement.

More recently there was the influx of fishermen from the crab and pollock fisheries. In 1970 Dutch Harbor–Unalaska was a fishing and cannery village in the Aleutian Islands archipelago with about 400 people. With the booming pollock industry it grew to 4,000 people by 1990. Dutch's population stabilized at about 4,000 inhabitants, who are spread across two islands. Thirty percent of its population is Asian, with growing numbers from Mexico and Somalia. Only about 8% are Native American. The median household income of Dutch Harbor is $70,000, compared with $50,000 for the rest of the United States.

The Aleutian Islands in summer are beautiful, emeralds in a sea of blue. In winter they are snow covered and harsh. Hills are white with black shadows. Snow doesn't layer on very thick before it gets blown off. The village of Unalaska lies in such splendor and remoteness. The nearest city is Anchorage, 800 miles and a three-hour airplane flight away. The feeling of remoteness is enhanced by turbulent skies above and the stormy sea surrounding the island. The Unangan people had a phrase for their home islands, "the shores where the sea breaks its back."[84]

When you drive into town from the airport, it isn't anything like you would imagine. The road is paved now, but lined with World War II pillboxes and bunkers, rusted cargo containers, abandoned trucks, and neatly stacked crab pots with black netting and red floats. You pass Amelia's Café, a latté stand, Safeway, Gas n' Go, Bering Sea Office and Electronics, Alaska Ship Supply, and then the Grand Aleutian Hotel. One gets a sense of impromptu living at first glance, a temporary place to make money and then abandon. You don't see many people walking. There is a lot of empty space between buildings. Net repair, engine repair, and electronic shops reside in wind-

*battered warehouses along the shore. Businesses seem randomly placed. Old cars and tires are recklessly abandoned along the gravel roads. Everyone drives a nearly new mud-covered pickup truck, destined for a short life.*

*But there is another aspect to the village, and that is over on the other side of the island. Across the Iliuliuk River, a short drive up a knoll leads you to a view of a picturesque village on the bay. In Unalaska there is an eclectic mix of houses. Some are tattered and wind-worn, unpainted driftwood brown. Others are painted in up to four different colors, like colorful gypsy wagons. And there are modern well-kept houses, especially on the surrounding hills. On the main street, there is the white Russian Orthodox Church with two blue domes topped by characteristic double crosses. At the end of Bayview lies the cemetery up on a hill overlooking the town. The weathered white markers formed of double crosses are planted in the ground and blown by winds to rest at random angles.*

*There is a deeper side to the community that a casual visitor can easily miss. One of the jewels in the city is the Museum of the Aleutians, a couple of small rooms, airy and filled with light, with a tasteful display of Aleut artifacts. Here one gets a thoughtful dose of what life must have been like before the white man arrived. There is evidence of an active community life, with a modern community center, senior center, and library. A skateboard park in the front yard of the city hall is a nice touch of mingling genera-tions. Community pride can be measured in the quality of its schools. The Unalaska High School is one of the best in Alaska. John Conwell, superin-tendent of the Unalaska City School District, proudly informed me that his school district earned prestigious recognition, including the following: Off-spring Magazine named Unalaska City School District as one of the top 100 districts in the country in 2000; US News and World Report selected Unalaska City School as a bronze medal winner in its January 2010 "Best High Schools" edition; and the Alaska Department of Education and Early Development nominated Unalaska City School as a 2011 National Blue Rib-bon School. Conwell reported that in 2010 the high school graduation rate was 96.4% and for 2011 the graduation rate was 95.7%. Graduates recently matriculated to Stanford University, Cornell University, Seattle University, Colorado School of Mines, New York University, American University, Uni-versity of Oregon, University of Washington, and many other colleges and universities throughout the country.*

· · · · ·

SIDEBAR 6: THE ELBOW ROOM, DUTCH HARBOR, ALASKA

*Dutch Harbor and the Elbow Room are usually mentioned in the same breath. The Elbow Room was symbolic of Dutch's reputation during the fishery's boom years. This landmark was once called the second most dangerous bar on the planet and was a legendary hangout for pollock and crab fishermen flush with cash.*[85] *The bar was known for its liquor, lines out the door, drugs and fights. "The bad reputation we had in those days was warranted," said Frank Kelty, former mayor of the town.*[86] *The tavern stayed open until 5 a.m.*

*In 1966 Larry Shaishaikoff and Carol Moller bought the Blue Fox Cocktail Lounge and the house across the street for $800, and renamed the bar the Elbow Room.*[87] *As the crab fishery grew in the mid-1970s, fishermen looked for an outlet to blow off steam. It's been said that rowdy behavior is a rite of passage for fishermen, and if there was a Mecca for these pilgrims it was the Elbow Room.*[88] *Bartender Ralph Harvey, a doorman who wore a striped referee jersey, reminisced that he once had eight fights in one night. He was able to break up four of them, and the other four caused a number of people to go to the medical clinic. The fishermen partied like they'd never see dry land again.*[89] *Another time a group of drunken crew members from a factory trawler on shore doused one of their "buddies" who'd passed out with rubbing alcohol and set him afire.*[90] *One citizen of Unalaska complained that the factory trawler people "drink beer in the bar and urinate in the streets."*[91]

*In the 1990s there was a movement to clean up the image of Dutch Harbor. Some citizens had seen enough sex, drugs, and rock and roll. The bar was renamed Latitudes a couple years before it closed. The rise and demise of the Elbow Room is symbolic of the boom years of the crab and pollock fishery, a life of danger, and the hard battles fought; its fall parallels the taming of the fishing industry.*

· · · · ·

# 11  Bridge over Troubled Water: Tranquility after the American Fisheries Act

*Science has to be independent of the political and social pressures coming at it from all sides.*                                      LEE ALVERSON

The American Fisheries Act of 1998 brought a sense of calmness back to the pollock industry. The madness of white-gold fever passed after Congress approved the measure. Some skippers now complain that the fishing is almost boring. The main impact of the American Fisheries Act was to end the race for fish, because each sector and vessel knew its quota allocation. That isn't to say that everyone has agreed with the process or with the distribution of allocations.

The end of the race for fish has been rewarding to those included in the allocation, and has had positive environmental benefits as well. The slowed pace of fishing powered the reduction of waste and bycatch. Yields and profitability have increased. Safety has improved. But there are still a few peas under the mattress. Bycatch of Chinook and chum salmon remains contentious. The effect of the fishery on the ecosystem is still an issue. The perceptions of fairness of the quota share and CDQ allocations, and the management council process, are other areas of concern.

## Rationalization

Linda Behnken represents a new wave of conservationist and politically active community-based fishers. She used to skipper her own boat, the *Morgan*, but now she shares a boat named the *Woodstock* with her husband and two sons.[1] They are a fishing family out of Sitka, Alaska. Behnken's other job is the executive director of the Alaska

Longline Fisherman's Association. She served nine long years on the NPFMC from 1992 to 2000, during the height of the onshore-offshore pollock fishery battles. This broad experience as both a harvester and a regulator has molded her view of the fishing industry.[2]

Behnken got a bachelor's degree from Dartmouth in English and environmental science and went on to get a master's degree in environmental science from Yale. She arrived in Alaska in 1982, and got a job on a family fishing boat. She bought her first boat in 1991. Now she longlines for halibut, sablefish, and rockfish, and trolls for salmon.

Behnken thinks that the AFA was one of the turning points of managing Alaskan groundfish resources, and not necessarily in a good way. She was on the NPFMC when the AFA became the law of the land in 1998. According to Behnken, the AFA process fundamentally changed the way management is done. In this case, the council process was bypassed and management was legislated. In doing so, the Magnuson-Stevens Act was violated, and the new rules were made by politicians rather than by council members with knowledge of the fishery. There was no public input, and it was arranged behind closed doors by those who would profit from the layout of the new playing field. She feels that because of the rationalization system, pollock and other fisheries are locked up in terms of policy by the limited set of players who are making a lot of money from a public resource. The particular system for pollock has tied up the quotas in corporations.

Although rationalization achieves the conservation goals of alleviating the race for fish, reducing bycatch, and improving safety, the socioeconomic impacts of limited entry and catch shares on Alaskan coastal communities have been profound. While some Alaskan communities have thrived under rationalization, others have lost access to local fisheries. The CDQ system also has had positive benefits, but it represents only 10% of the fishery and much of that is leased to the major companies.

One of Behnken's main concerns is that quota systems will cause "the move toward corporate control and ownership of the fishery."[3] She believes that protections need to be in place so that fisheries are a thriving part of the community and to help keep the remote communities of Alaska alive. Behnken is also concerned with the push to create broad-based marine protected areas, and that "environmental organizations wanting to close down large areas, creating more marine wildernesses, instead of finding out about the different impacts of different gear types," will give an advantage to large vessels without contributing to long-term marine conservation. She believes sustainable fisheries and coastal communities are linked, since coastal residents are advocates for marine conservation. She would prefer to see a stronger small-boat, community-based fishery.

Behnken isn't alone in her concerns. One fisherman said of the quota system, "I want to go out there and be competitive." An industry representative responded, "A lot of people just don't like the idea of change. They're not ready to see the last of the buffalo hunters tied up in a rational system. Those are the real Alaskans. They like rolling the dice out there, going out not knowing how much they're going to get."[4]

The Environmental Defense Fund (EDF) is on the other side of the rationalization fence. About twenty years ago, scientists at the EDF analyzed what was going wrong in the world's oceans. They identified several problems, including the loss of biodiversity. They believed that overfishing was the chief cause of biodiversity loss in the short term, and that habitat loss and pollution are key factors in the long term. EDF established its Oceans Program in 1989 to tackle these problems.[5]

Dr. Rod Fujita was one of the founders of EDF's Oceans Program. Fujita is another new brand of environmentalist, a scientist who turned to more active participation in conserving the oceans. Fujita went to Pitzer College and got his PhD at the Boston University–Marine Biological Laboratory joint program in Woods Hole. He has worked in marine science for over thirty years. He is a gifted communicator and knowledgeable scientist. His book, *Heal the Ocean*, lays out present problems facing the world's oceans and maps out his view of solutions that include creating marine reserves, supporting sustainably caught seafood, and supporting quota share programs for fishers.

Fujita says that EDF doesn't compromise on the environment and likes to say they "are strict on goals but flexible with people." "We believe in the creativity and ingenuity of people to solve problems given a clear set of goals." According to Fujita, identifying the goals is critical. "Is the goal to end all trawling? Or even to end all fishing? Or are the real environmental issues to end overfishing, reduce bycatch, and prevent habitat destruction? From the industry point of view, do they want to simply continue the status quo? Do they need to keep taking more and more fish? Or is their true interest in continuing to fish, and to maximize the value of the fish that they catch?" Fujita says that EDF believes in creating situations where dialog can happen between competing interests so that people can get off their positions, reveal their true interests, and work together to find win-win situations.

Fujita and EDF were early advocates of rationalization programs. According to Fujita, rather than focusing on the symptoms of overfishing, bycatch, and habitat destruction, they advocated solutions that get to the heart of the matter—economic incentives for overcapitalization. Catch shares seemed like the best way to solve dual goals: fishermen want to maximize value and environmentalists want to prevent overfishing. Fujita notes that not all catch share programs are perfect, but mechanisms can be built into them to make them work for different situations and communities.

Fujita says that fishermen "often describe themselves as being engaged in a noble profession, and of course, they are." They work hard and risk life and limb to feed the nation wholesome seafood. But according to Fujita, fishing is not a right, but rather it is a privilege granted to them to catch and sell a public resource. "They don't own the fish, but since they are using a public resource, they have a responsibility for being good stewards of it. Catch shares allow fishermen to align their fishing practices with conservation and with their own values—few if any fishermen enjoy discarding fish or chewing up bottom habitat."

"EDF strives to find ways to achieve stringent environmental goals that accommodate industry interests by unleashing creativity and good intentions on all sides." Fujita now directs EDF's Ocean Innovations group, a think tank dedicated to developing creative solutions to major ocean conservation problems. He says that this perspective and the expertise they have accumulated have helped them work with fishermen and government officials to solve major fishery challenges in New England, the Pacific Coast, and the Gulf of Mexico, and increasingly in other countries such as Mexico, Belize, Cuba, and the European Union.

There is often an evolution in the relationship between NGOs and industry. It usually starts as a conflict in an oppositional context and then escalates. The escalation is sometimes the result of the council process, where people talk in sequence in a microphone at the hearing, instead of having a dialog. Things get defensive. The council meetings often feel like "public spearings" instead of public hearings. A better way is to find new paths to develop common objectives, where parties find common goals and new leaders emerge.

The tragedy of the commons is often cited as the need to rationalize fisheries. This is the principle that when there is a common resource pool, individuals will look after themselves and deplete the resource even though everyone recognizes that it isn't in their self-interest to do so. But the tragedy of the commons is a metaphor that simplifies the complexity of the problem. There are actually two issues: too many fish being caught and too many boats.[6]

The term "rationalization" comes from the concept of the rational fisherman. The idea is that if a fisherman knows that he controls the future of his resource, he will be reasonable and will fish it in a sustainable manner. The concept of rationalization or catch shares essentially privatizes a public resource, which is controversial. Supporters of catch share programs, like EDF, point out that the resource isn't owned, but the privilege to exploit it is. However, others contend that if you have the right to exploit it, rent or lease it, or even sell it, then it sounds and smells like ownership. Once a precedent is established, it is difficult to say how the courts will interpret legal challenges of the "owners" against regulators.

There are layers of complexity and degrees of shading embedded in the concept of catch shares. It is an issue that involves areas of human endeavor from economics to philosophy and political science. It pits environmentalists against environmentalists, academics against academics, politicians against politicians, and fishermen against fishermen, and all possible combinations of the above. Rationalization has caused a rift within the environmental community. While EDF promoted rationalization by focusing on an economic approach to control industrialized fishing, Greenpeace took the "factories of doom" approach and disapproved of handing over a valuable gift to an already politically muscular group of owners.

In principle, the rationalized fishery is closed to new entrants and the quota gets distributed to a closed class of owners, privatizing access to the fish.[7] The federal government takes what is a public good, and gives the privilege to harvest it to a small defined group of individuals and companies. This has incensed those who have been excluded from access.

From the point of view of the fish, the problem of open access fishing is not a problem for the resource. The problem is not who catches them, but how many are caught. Who catches the fish, and how many participate in the activity, *are* problems for the fishing industry in terms of "economic growth" and "economic returns." From the perspective of large industries and the politicians that they lobby, healthy economic returns are those needed to expand and grow, and to invest in the political process.

Rationalization is a way to manage the fishermen and not the resource. Claims that catch share systems can eliminate overfishing are overstated. What the catch share system does is control the competitiveness of the fishery and increase the economic efficiency of participants. A strong limit on the number of fish caught and the will to maintain those limits are what prevents overfishing of pollock.

There may be some ecological benefits to rationalization, such as increased utilization and reduced bycatch. Regulation also attains these goals. There may also be unpredictable ecological effects from either catch share allocations or regulation, such as a shift in the location and season of fishing, which could have positive or negative effects on the ecosystem.

In theory and in practice, the creation of enhanced individual fishing rights opens opportunities for fishing companies to reduce costs and to increase the market value of their catch.[8] The value of the resource is realized efficiently. However, the value of the resource, or as economists call it, the *rent*, is held by fewer owners rather than being spread over a larger community. In the first case, the profit can benefit shareholders, and assist political action committees and other entities to maintain the status quo. In the latter, and "less efficient" industry, there is less profit and more of the value of the resource is spread into the community to support more fishermen and the infrastructure to maintain more boats.

Some environmental organizations realized the threat of rationalization. They could see the danger in big companies taking over fisheries. On the other hand, it has delighted those who were given access, some environmental groups who see this as the only practical way to manage access, and many fisheries managers. From the point of view of the fisherman who has quota as a consequence of his catch history, why shouldn't he be rewarded for his investment?

From an ideological point of view, the IFQ system has some interesting contradictions. Economic liberalism (most often associated with political conservatives and libertarians) promotes a laissez-faire approach to government management of the economy, but with some state action to improve individual and social conditions. For so-called neoliberalists, the assumption of societal and ethical responsibility is released, as is state intervention in economics, because it has perceived negative effects on economy and society.[9]

Neoliberalists link free markets with political and personal freedom. On the other hand, neoliberal proponents of "rights-based" fisheries management assert that resource management cannot move beyond the tragedy of the commons until property rights are assigned. The contradiction here is that the "free market" did not evolve, but instead it is set up, regulated, and controlled by government action.

During heated debates in the Senate over the reauthorization of the FCMA and passage of the Sustainable Fisheries Act in 1996, the chamber agreed to a four-year ban on IFQs (or catch shares) until a study by the National Academy of Sciences could be completed.[10] Senator Stevens and some of the inshore sector were at first opposed to IFQs. Most of the smaller-scale fishermen and environmental groups also opposed the concept of rationalization, fearing a corporate takeover. Guaranteed catch shares violated one of the traditional ideas of fishing: if you are a good fisherman, you catch more fish and earn more money. Catch shares make the fishermen more like crop harvesters.[11]

In 1997 the Alaska delegation on the council successfully blocked further discussion of rationalization. However, by this time the Washington delegation to the Senate and the at-sea processors were in favor of it. The factory trawlers desired some long-term stability in their quotas, realized that they were overcapitalized, and needed to limit entry to the fishery. They were getting squeezed out of the fishery by the Alaska delegation on the NPFMC. Most economists I've read and talked to were also in favor of some form of limited entry through catch shares. Otherwise the huge economic value of the fishery "is very likely to be squandered by overcapitalization."[12]

Although people are careful to say that rationalization is not private ownership of the resource, but a privilege to use the resource, it is clear that others don't share this nuanced view. The difference is that property rights are permanent, compensable, and tradable, and not temporary or revocable. Given that it will "take an act of Congress" to change the pollock quota system and the political process, the right to fish pollock is not temporary and hardly revocable. Clearly, many in the industry view the fishery as privately held. Clem Tillion, a friend of Ted Stevens and known as the "fisheries czar" at the time of the AFA, was one of the main proponents of catch shares (or ITQs). He said, "People don't like ITQs, but I happen to like them because I believe that private ownership is better for management than public ownership."[13]

The system of "cooperatives" that developed for pollock in the Bering Sea created private company shares of the total allowable catch by agreement. The holders of limited entry permits "cooperatively" agree on a harvest sharing system.

Several questions arise, given that rights-based fisheries are profiting. Why continue to use public funds to support fisheries research and enforcement and the decision-making process?[14] Federal subsidies to the US fishing industry are estimated at $6.4 billion from 1996 to 2004, not including funds for fisheries management, port construction, and maintenance.[15] Subsidies during that period were worth about $420 million to fisheries in Alaska. And why are the rights handed down in perpetuity rather than for the generation time of the target fish when they are assigned?

The question of rationalization of the fisheries is one that stretches back to the time of Hugo Grotius. We haven't been able to answer the question of who owns the resources of the sea in the past 350 years. Clearly there needs to be some form of limited entry to keep a handle on the amount of effort that is applied. Almost fifty years ago, Graham's Great Law of Fishing said "Fisheries that are unlimited become unprofitable." Somehow there must be a balance in how rationalization is set up to meet the needs of industry and community-based fisheries, permitting new fishermen to enter the fishery while protecting the interests of the experienced fisherman. Again, Michael Graham said almost fifty years ago, "There must sooner or later be control. It is a difficult problem—how to institute control, and leave this industry freedom, without which it will die of boredom." He continued to say, "Give privileged places to men who land inshore fish, to skipper-owners, and perhaps to small managing owners," and further, "Any system starts by recognizing the relative magnitudes of existing firms in the business, thereby giving permanence to a success that may only deserve to be temporary. Let us reduce that evil as much as possible: instead of keeping out new men, let it be ruled that anyone can win a place by raising existing standards. . . . Because control inevitably makes for more security, let us make less wherever possible, as by encouraging managing owners, discouraging limited liability companies."

## Community Development Quotas

In 1990, 25% of the population in villages along the coast of western Alaska was living below the poverty level. There were social problems in the populace of about 23,000. The villages were not linked by roads or connected to a major population center. Commodities and people traveled by air or ship, making the cost of living high. In some regions, the unemployment level reached 43%.

One of the more progressive actions of the pollock fishery has been the formation of community development quotas. The idea of CDQs first came up in the 1980s when the NPFMC provided a small amount of Bering Sea fish, mainly halibut, to local vessels. This was a welcomed boost because impoverished western Alaska community groups didn't have the capital to invest in groundfish fisheries.[16] Then in 1989, Senator Stevens introduced legislation in Congress to create a CDQ to increase the stake of coastal Alaskans in the groundfish fisheries, but it wasn't enacted.

In 1992, the CDQ passed as an amendment to the NPFMC during the onshore/offshore squabble. Some said that the villagers would squander the money.[17] The CDQs are shared among six community coalitions. Under the original CDQ amendment, these six Alaska-only nonprofit organizations would use a grant of quota from Bering Sea fisheries to help disadvantaged coastal communities. The organizations centered around sixty-five eligible villages in western Alaska with 27,000 people. Each regional association would associate with a regional fishing partner (i.e., company) and submit a proposal to the Governor of Alaska outlining a development plan describing what they would do with earnings for a specific quota of pollock. Each company would have to demonstrate how it uses profits to help impoverished western Alaska villages. The proposals would be reviewed and approved by the Governor and Secretary of Commerce. The fishing partner would then acquire that year's quota. There were restrictions placed on what the regional association could do with the profits.

Although the goals are lofty, the CDQ hasn't been free of problems. Over the years, the program has been nagged by fights over quotas, state and federal scrutiny of CDQ business affairs, contention over state versus federal oversight of the program, and questions over legitimate use of CDQ profits.[18]

In 1996 the CDQ program became law as part of the Sustainable Fisheries Act. The new law required that the communities benefiting needed to be within 50 miles of the shore and they must consist of residents conducting more than one-half of their commercial or subsistence fishing in the Bering Sea or Aleutian Islands. With the passage of the AFA in 1998, the CDQ amount was raised from 7.5% to 10%.

In 2003, the combined assets of the six CDQ companies were estimated at $227 million. Competition among the companies for quota had become combative.[19] Three of the companies lodged complaints with federal officials about allocations by the state. Legislation was introduced by Senator Stevens as a rider on the fiscal 2004 budget to add the Aleut Corporation to the CDQ program. The deal would be worth about $30 million. Critics complained that Stevens' son Ben sat on the board of the Aleut Enterprise Corp., a subsidiary of Aleut Corp. Stevens responded that he didn't know his son was on the board. The Aleut Corp. would represent the village of Adak, with a permanent population of seventy-five people.[20] Adak was never permitted to become a CDQ entity, but instead a portion of the crab quota was set aside for village use, and a special allocation of pollock for the Aleutian Islands has been given.

In 2005 the worth of assets held by the CDQ groups was $350 million. State officials were unhappy with the federal government usurping control over the program.[21] The allocation fights among the CDQ groups were also causing uncertainty in the partnered industry groups because they didn't know how much fish they would be receiving.

In 2006 Congress amended the Magnuson-Stevens Act to incorporate changes in the CDQ. This included a ten-year cycle in allocations among the six CDQ groups and a provision that up to 20% of CDQ group's investments could be non–fishery-related. The state would now review each CDQ for their overall record and modify allocations by a 10% reduction for all or part of the next ten-year cycle. Each CDQ group would now develop and approve its own community development plan.

The CDQ groups have tax-exempt status, which gives them a huge advantage in the fishery. In 2010 four of the six CDQ groups spent $135,000 on lobbyists to push for tax-exempt status of their for-profit subsidiaries as well. Those firms and their political action committees contributed $36,000 to Senator Lisa Murkowski's re-election campaign.[22]

By 2010 the total assets of the CDQ groups were around $900 million. The poverty rates reported by the associations from US Census data of 2000, compared with the average of the years 2005 to 2009 (from the American Community Service), indicate that poverty rates went down for two of the six CDQ groups and up for the other four. Overall the proportion of people living below the poverty level slightly increased from 25% to 26% between US Census years 1990 and 2000. The most recent records have not been published. However, some rough comparisons can be made from county-wide poverty rates reported in the 2010 census.[23] Most of the western coastal Alaska counties had poverty levels of 9% to 31%. Overall, except for the Aleutian Pribilof Island Community Development Association area where rates are about equal to that of the rest of the state of Alaska, association area rates still appear to be 1.5 to 3 times higher than the general rate for the state. Mean personal income of each CDQ group almost doubled from 1999 to 2007, slightly higher or similar to state-wide increases, and employment in most groups increased.[24]

There has been a recent report that executives of some CDQ groups, nonprofit entities which are entrusted with a public asset, are making large salaries. The executive director of one group was reported to have been paid nearly $1 million in 2008.[25]

Through their investment in the fishery, the CDQ groups are said to now control 40%–45% of the pollock trawl fleet in the Bering Sea.[26] One representative of a major company asked me, "How much do you think the trawlers are worth?" I responded, "Probably anywhere from

$10 million to $40 million." He answered back, "They aren't worth anything. All the value is in the quotas." One might ask, what happens if
the population declines? If the boats aren't worth anything without
the quota, and the quota is really an item to sell and lease, then if that
value diminishes, whoever is left holding the bag has an empty one.
The CDQ seems primed to become a major battleground in the pollock
industry. Bob Stors, then vice president of the Unalaska Native Fishermen's Association, said in 2000, "Nine thousand years of participation
in the fishing industry have been recorded in the community of Unalaska, Alaska. The idea that this can be usurped by well-financed, powerful industry interests is disgusting. There is a lack of recognition
that small boats are part of the future. Fishing-dependent communities are more conservation-oriented than the large corporations because communities are concerned with sustaining the participation in
the fishery over the long term."[27] But all that seems to be changing. The
revenue from the CDQ fisheries is worth about $150 million–$175 million each year.[28] And the investment isn't primarily in small boats. The
newly found wealth of the CDQ groups is being used to make plans for
a new home port for the Seattle-based fishing fleet in Alaska. A lot of
that fleet is now owned by CDQ groups. The Seattle side states, "When
you look at the advantages, they [the Alaska nonprofits] obviously are
going to end up being king of the hill. We need a more equal playing
field." An executive of one of the CDQ groups countered, "I think this
is revisionist history, the letter [written by the Seattle group] seeks to
recolonize Alaska."[29]

## The Council Process

One of the major consequences of the FCMA was to establish the eight
regional councils to oversee management of the nation's fisheries.
The North Pacific Fishery Management Council has jurisdiction over
900,000 square miles of ocean off Alaska. It has eleven voting members: six from Alaska, three from Washington, one from Oregon, and
a federal representative, the Alaska Regional Director of NMFS. There
are nonvoting members who represent the state fisheries agencies,
commercial and recreational fisheries, US Coast Guard, US Fish and
Wildlife Service, US Department of State, the Pacific States Marine

Fisheries Commission, and the public. There is an Advisory Panel made of industry representatives, and a Scientific and Statistical Committee made of biologists, economists, and social scientists.

Ross Anderson of the *Seattle Times* wrote of the NPFMC, "Congress in turn defers most of the decisions to a narrow-minded, politically loaded federal council, made up mostly of people who work for fishing companies that stand to make millions from those decisions."[30]

The amount of money at stake in NPFMC decisions is enormous. Since the council is dominated by industry representatives, there have been accusations of conflicts of interest. "The United States' richest fishery is controlled by a federal council so riddled with conflicts of interest that its actions result in millions of dollars in benefits to some council members or their companies. . . . The council emerged in public view as a sort of Tammany Hall of the Pacific."[31] Pollock fisherman Dave Fraser once said, "The allocation [of pollock] is done by the council, which has proven itself incapable of disregarding its myriad political and economic conflicts."[32]

With his usual refreshing candor, Wally Pereyra, a council member at the time, said, "The politics are so blatant it forces you as a protective measure to become part of the problem. It's a corrupting process."[33] Of the eleven NPFMC members at the time of this writing, four owned fishing businesses, two were industry employees, and one was a an industry consultant. Dave Fraser, who was at the time skipper and owner of the *Muir Milach*, said, "People talk about the fox guarding the hen house. But what we've got here is a few roosters divvying up the hens."[34]

Why were the councils set up to favor the industry? In 1976, Bud Walsh worked with Senator Warren Magnuson of Washington to write the FCMA and afterwards to set up the management councils as mandated by it. He said that the councils were established under the control of industry because of mistrust of scientists, who might set catch limits too low.[35] Even people who helped draft the law consider that the process doesn't work because of conflicting interests of the fishing industry members. Don Bevan, a former director of the University of Washington's School of Fisheries, said, "It just isn't good public policy having people with a direct financial interest making decisions on those financial interests. We wouldn't stand still for a minute with a city council run by contractors making decisions on city contracts."

Lee Alverson said, "We're talking about billions of dollars that people can shift around between industry groups, and they're going to vote for themselves."[36] Alverson continued, "Fishery management failure is weeded in institutional paralysis and political involvement. The regional fishery management council process is weak in that it's open to considerable influence by particular special interest groups."[37] The FCMA exempted members from usual conflict of interest regulations beyond disclosure and possible recusal from participation.[38]

Another former council member told me, "It is hard to see a change will be forthcoming because the power structure in the industry and council is intertwined, and is politically well connected. It has become a big industry, too big to let die." That person continued, "The council members who supported the catch shares system in 1998 were committed to maintaining the economic sustainability of the pollock resource in which they had a vested interest, and supported a system which protected those interests."

The ethics instructions of NPFMC to council members state, "With disclosure, you may fully participate as a Council member in a matter in which you have an interest in a harvesting, processing, lobbying, advocacy, or marketing activity." The exceptions are if there is greater than 10% interest in the total harvest, or in the processing or marketing of the total harvest; or there is greater than 10% ownership in the vessels within the fishery. No one would be excluded from making decisions about the pollock fishery under these conditions. Further rulings have been passed reinforcing disclosure rules, and exempting council members from normal federal conflict of interest laws. Recusal from voting is voluntary.[39]

However, these negative aspects being said, in thirty years the council has a proud history of never having overridden the allowable biological catch (ABC) put forth by the advisory Scientific and Statistical Committee.[40] The main controversies have come from how the quota has been allocated among user groups.

## Ecosystem Effects

Ecosystem issues confronting the pollock industry include the continuing conflicts involving Steller sea lions, fur seals, and seabirds. The real problem of ecosystems and fisheries management is the con-

flicting goals of the two. According to the theory of fisheries management that is based on a single species, the goal is to reduce the stock
biomass by 50% or more in order to maximize productivity. Some fisheries scientists have even indicated that reducing a population to 10%
of its biomass may give good sustainable yields.[41] Ecosystem-based
management is an integrated approach that considers all activities
that impact ecosystems as a whole, with the goal of sustaining the services the ecosystem provides. Reducing the biomass of a major prey
species like pollock to 50% or less of its biomass is going to impact
other species. The other natural populations that feed on pollock have
adjusted their own abundance levels to the density of pollock before
regulation, and can't dial back on their take without impacting their
own vital rates. They have to switch to alternative prey. But if those
prey are also the object of fishing, then there is a problem. The result
of commercial fishing along with natural predation is a huge demand
on the ecosystem.

Of the pollock stocks that experienced marked declines, if not collapses, including the western Bering Sea, Gulf of Alaska, Aleutian
Islands, Donut Hole, and Bogoslof stocks, all experienced increasing
harvest levels for a period of years when the stocks were at high levels.
Like the number of fishing boats, predator populations also increased
in response to high prey levels. When the stocks began to decline, the
high commercial harvest levels continued because management didn't
see the declines for a number of years. Likewise, the predators can't
reduce their own metabolic requirements, and the declining numbers of pollock as prey were exposed to very high levels of natural
mortality. These enhanced rates of natural mortality aren't accounted
for in the underlying assumptions of MSY theory. Furthermore, the
fishery targets the larger animals in the population and the predator
community eats the smaller animals. The result is a double whammy
and collapse is hard to avoid. In a recent book, Fowler and McClusky
wrote, "There is more to the complexity of natural systems than can be
expressed by population models."[42] They go on to say that "most of the
progress that has been made in considering greater levels of complexity is mostly rhetoric."

Fowler and McClusky advanced an idea to position human harvests
to fall within the normal range of natural variation seen among other

species. This is founded on the concept that patterns seen in natural systems have evolved to account for the complexity in them and the unknown properties and interactions. The total mortality that a species experiences in nature is already adapted for its own sustainability, or buffered against the system, and once we exceed that we are treading upon the natural abilities of the population to sustain increased mortality. When an environment turns hostile for survival and the population begins a downturn, the interaction of an increasing demand by ecosystem predators and increasing harvest levels can pose a significant hurdle to stability.

In the case of pollock, the fishery takes twenty-eight times as much as the amount taken by any species of marine mammals. The take is nearly five times greater than the combined takes of all six mammal species considered. Therefore, according to Fowler and McClusky's theory, the human harvest is a statistical outlier falling outside the range of natural variation. Furthermore, ecosystem-wide fisheries removals, of all species, from the eastern Bering Sea is about twenty-seven times greater than the total mean consumption rates of twenty-one marine mammal species. For the harvest of pollock to fall to the mean take of marine mammal species, the harvest would have to decrease from about one million tons to 200 thousand tons per year. If the industry is worth $1 billion, the drop in revenue would be $800 million. That would be a practical problem. Ted Ames, a fisherman, fisheries researcher, and MacArthur Fellow, was asked what he thought about the behavior of large companies in Alaska holding permits to fish pollock in relation to the industry's need for a continuous supply of fish. He offered a quote from a friend, "Remember one thing, never get between a fat hog and a trough. He'll run you over every time."[43]

There are paradoxes involving single-species population dynamics in the larger context of ecosystems. Isolating a single species and considering the effects of management actions upon it can be misleading. Any changes that occur are going to also impact the larger ecosystem. Complexity and uncertainty in the system that is being managed can result in unintended consequences.[44] On the other hand, it is hard to gain focus on an ecosystem. Often such a holistic approach is taken to the exclusion of the basic factors affecting population dynamics. Probably at the current state of knowledge, the best we can do is to have an understanding of the complexity and the system of interactions among organisms in the ecosystem that might be impacted by our own intervention.[45]

## The Continuing Environment-Industry Struggle

Greenpeace has taken a more aggressive stance on the fishery than some of the other nongovernmental organizations (NGOs). Many of the environmental groups have overlapping but distinct interests. Since they depend on fund-raising for their existence, to avoid direct competition, the NGOs have evolved different niches within the conservation community. A group like Greenpeace has come to be known as an activist group, and they depend on calling public attention to their campaigns. In part, this has evolved as a countermeasure to balance the influence of the well-funded opposition in the political process. When you are dealing with a hot political issue, whatever tools are available are used to gain the public's imagination and to gain advantage. This counterbalances the role of the industry's financial power in the political process.

Other environmental groups take a more cooperative role to work with industry. The World Wildlife Fund works with industry to support corrective action such as the Marine Stewardship Council (MSC) certification program. The MSC sets environmental standards for sustainable harvesting. The Environmental Defense Fund aims to find solutions on specific issues. They are leaders in the charge for rationalization of the fishery and creating marine reserves. The Center for Marine Conservation specializes in small fisheries conservation and preservation of sea turtles and marine mammals. The Audubon Society specifically works on birds, and the Nature Conservancy on refuges and bycatch. The Nature Conservancy, interestingly, owns some quota for the West Coast groundfish stocks. Like Greenpeace, Oceana is an activist group that stirs the pot in many marine issues and captures the imagination of the public. The Marine Fish Conservation Network is a body organized to represent several groups to influence federal fisheries policy. They act as a watchdog cooperative that reports on the successes and failures of federal policy. The Network has grown to represent over 100 environmental and conservation organizations, as well as progressive fisheries associations. Celebrities such as Katherine Hepburn, Ted Danson, and others have signed on to promote marine conservation. Private foundations, such as the Charles and Betty Moore Foundation, the Pew Charitable Trusts, and the David and Lucile Packard Foundation, have also gotten involved in marine fisheries issues.

On the industry side there is a plethora of associations representing their interests in the public media and politically as well. Some of those are the National Fisheries Institute, the At-Sea Processors Association, United Catcher Boats, Groundfish Forum, Central Bering Sea Fishermen's Association, Pollock Conservation Cooperative, Marine Conservation Alliance, Sea Alliance, United Fishermen of Alaska, Alaska Whitefish Trawlers Association, Yukon Delta Fisheries Development Association, Pacific Seafood Processors Association, Yukon River Drainage Fisheries Association, and Alaska Longline Fishermen's Association.

What has evolved in the industry-environmental struggle is a system of checks and balance. One leader of the Norwegian fishing community, Tor Tollessen, told me that he sees the benefit for groups like Greenpeace to counterbalance the industry, but some common sense is needed. He talked about the Norwegian term *bondevet*, literally "farm sense." In other words, one needs to have common sense, to know how things work, and to keep a balanced perspective, rather than to irritate an already tense situation. Fisherman Dave Fraser made a similar observation, that the environmental groups were a needed counterweight to human greed. Lee Alverson suggested that the environmental movement added a new perspective to the science, such as genetics, biodiversity, and spatial management. They raised new questions, and pointed out that you can't make simple arguments based on simple models of population dynamics.

At the high point of the factory trawler protests in 1998, NMFS assessed the health of US fishing grounds and concluded that the North Pacific was the only ocean bordering the United States where the fish stocks were not overfished. Especially in the Bering Sea, the pollock stocks were pronounced healthy.[46] Many environmentalists and scientists alike argue that the ocean system is complex and removal of millions of tons of fish has a lot of uncertainty with regard to effects on the ecosystem. Getting back to Greenpeace's comparison with the forest, you can harvest fish populations at sustainable levels, but those levels of harvest may destroy the ecosystem.

There was a common theme among several environmentalists that I interviewed. They said that although the eastern Bering Sea pollock stock seems in pretty good shape, the ecosystem impacts of the fishery are still unknown.

They believe that the management agency adequately surveys the resource, they have good information, and the industry has worked hard to attain MSC certification. However, there is concern that the pollock fishery is put forward as a model for fisheries management, when it is really a special situation and not a model that is necessarily transferable. In other fisheries, they are still fighting overcapitalization, trying to put observers onboard, and struggling with bycatch and bottom habitat destruction by trawl nets. The pollock fishery has already cracked these nuts.

Harvesting of pollock has its own problems that need to be confronted, such as the spatial distribution of harvesting and how that may impact other components of the ecosystem, such as mammals. Environmentalists suggest that a more sophisticated approach is required where we consider the foraging needs of predators. For example, sea lions have a high-energy-demand lifestyle. They don't store energy like some other species, so they can't sustain periods of starvation. A fishery based on the philosophy of reducing the density of the pollock to half or even less of their prefishing levels is going to impact the animals that depend on pollock as prey. Certainly, seasonal depletion in some "hotspot" fishing areas where mammals and prey aggregate is going to be serious.

### Bycatch Issues

Bycatch of Chinook and chum salmon in the pollock fishery has been an issue for many years. There was strict bycatch enforcement based on limits from 1982 to 1990, but when the pollock fishery was Americanized, bycatch reduction measures were reduced and usually not even enforced. Beginning in 1994, area closures were put in place to limit fishing in known hotspots for chum and Chinook salmon to reduce salmon bycatch under pressure from the salmon fishing industry.

Since 2006 there has been an industry-run rolling hotspot closure. Bycatch is monitored by observers and compiled by a private company, and areas of high salmon catches are closed. This seems to be effective, but in 2007 a record high catch of 120,000 Chinook salmon and 700,000 chum salmon was taken. Some of the salmon bycatch is believed to come from endangered salmon runs, and it will be of interest to see how the ESA plays into the mix in the future.

Western Alaska salmon fishing communities are pushing for reductions in the hard cap on Chinook and for the institution of a hard cap on chum salmon. There is a competitive interaction among the salmon and pollock fishers. Because of the CDQ associations and the wealth derived from the harvest of groundfishes, this is causing acrimony in the community. Do they want more salmon, which support more local western Alaska community fishermen and is a cultural tradition, or do they want more groundfish, which is providing wealth to these communities but is an industry where they have little direct participation?

## Management of the Fishery

The state and federal scientists who are responsible for recommending catch quotas put their necks out on the chopping block every year. Stock assessment and management biologists are talented, well trained, and well intentioned. But can we really control the productivity of natural populations and understand the consequences of harvesting a large portion of the stock biomass simply based on the output of mathematical models? Sidney Holt reportedly said that aspects of the models routinely used by biologists actually induce overfishing.[47] He added that rather than aiming high to stimulate fish populations to grow, we should be aiming low, for a catch that is well within the biological limitations of the species, the possible fluctuations of its environments, and possible errors in estimations. Rod Fujita of the Environmental Defense Fund said, "Evaluation of marine environmental policy should go beyond economic indicators such as profit margin, total revenue generated, and maximum sustainable yield targets. Ecological indicators of the status of marine systems are very important."[48]

Can we accurately determine the number of fish in the ocean, and then determine the maximum number of them we can safely remove? Do we really understand the nature of what we are studying? Can we engineer the abundance of fish populations?

Alan Longhurst said that "fisheries science is usually perceived by its practitioners as being a critical and quantitative activity, deeply dependent on mathematical analyses." He goes on to explain that this statement demonstrates what went wrong with the science: it forgot that it is heavily dependent on two other disciplines—biology and ecology.

The unique natural histories that fishes have evolved over millions of years make a difference in their response to fishing and how we should manage them. The really hard question is how to implement that information. Accounting for the individual differences among species makes management of fisheries a very difficult job. In his blockbuster anti-industrial fishing treatise, Charles Clover said, "Conserving fishes is one of the world's most challenging intellectual questions."[49] This is a surprising statement. Across the vast spectrum of difficult problems facing the world, who would have guessed that fisheries science is one of the most challenging, or for that matter, even intellectual, questions? But it is difficult, and made even more so by the complex economic, social, and political questions involved in making public policy decisions.

Are Alaska pollock stocks going to decline? The former director of the Alaska Fisheries Science Center, Bill Aron, said of the future of the pollock stocks, "Oh, it will decline. No matter how good your management is, animal populations undergo normal cyclical changes that will overwhelm any ability to manage."[50] His predecessor, Lee Alverson, quipped, "If I had to forecast, I'd say that one of these days the pollock stock will go down. Every one of these stocks has a cyclic decline." In a recent paper, a group of scientists, including the lead pollock stock assessment analyst, predicted that sometime in the next fifty years pollock will decline due to global warming.[51]

This prediction is likely to be a safe bet considering that the history of populations is to fluctuate cyclically owing to many interacting factors. But whether collapses are due to overfishing or global warming is an argument that scientists throughout history have never been able to resolve. There are still fierce debates about whether climate, fisheries, or their interaction causes population collapse. Similar themes are repeated in the world's oceans over and over again.

Clover in 2006 said that the central problem for fisheries science is counting fish. Inaccurate numbers might be one reason for desperate conditions in North Sea stocks in the 1980s and 1990s. Clover also cites several other issues, including the overriding of scientists' recommendations by the agendas of politicians or fishermen. In the case of pollock, in the last two years many fishermen have advocated a more conservative position than the Scientific and Statistical Committee (SSC) of the NPFMC, which is an interesting turnaround. Clover posits that scientists may be just another special interest group in the process, with only a marginal interest in the conservation of fish, on their way to career advancement. This is an accusation that stings, and may be the rare case, although I don't think he is alone in thinking of that possibility.

In the case of Alaska pollock, harvest levels seem to be moderate and ABCs have never been exceeded. We have pretty good knowledge of fish stock abundance through surveys. But there are many unknown factors, such as how many self-sustaining reproductive populations are in the mix, and how they are accounted for in the assessment. Furthermore, what is the effect of a large, continuous, and selective fishery on the evolution of life history traits? Then there is the contentious issue of how many fish can be removed without adversely impacting the environment that sustains them, or their own capacity to reproduce.

Even without fisheries, populations go through natural cycles of abundance, increasing and decreasing within the boundaries of their environment. Natural populations have adapted to the high and variable mortality in early life caused by the weather, food variability, predators, and disease. When conditions are very favorable, the population increases until it reaches the carrying capacity of its environment and then some substrate becomes limiting. That may include space or food availability. Eventually predators or disease can impact the population and it may decrease. Often the population and predators or incidence of disease have cycles that are inverse in phase. When conditions are unsuitable and the limits of adaptability of the population have been stretched too far, it may collapse.

The force of a fishery counters the natural adaptation of the population by increasing mortality of the oldest ages in the population. By reducing the age spread of the population, it causes more dependence for the health of the population on successful recruitment, making it more vulnerable to an environmentally driven collapse.

What seems to happen under the conditions of heavy exploitation for many marine fishes is that the resilience of the population is altered.[52] Natural populations have evolved an age span to adapt to variations in recruitment. A long age span makes up for periodic recruitment episodes that support the population. Long age guarantees that individuals will be present to mature and spawn, even if the offspring from most years have poor survival.

For Alaska pollock, the maximum age is about thirty years. Such a long life provides a way to hedge bets against highly variable survival during early life. With such a strategy the population can maintain itself through a decade of low recruitment. But now they have a realistic life of about eight years. The population is built on a foundation of fewer age classes. Since 2009, the bulk of the fish supporting the population are less than 5 years old. The population is now dependent on the strength of incoming year classes.

Another problem with modern fisheries has been the difficulty and reluctance of management in dealing with declines in fisheries. In the case of pollock, when stocks had a period of growth, quotas were increased. But just like the fishery, predator populations also responded to increasing prey levels with their own increasing abundance. The problem arises when the stocks begin to decline. Usually the declining trend isn't caught until it is a couple years into the trend. It has been only after the fact that we realize that catches were increasing on a declining stock. The pollock gets hit at both ends: high natural mortality by predators on young fish and high mortality on older fish by the fishery. One solution is a reliable forecast, which we have seen is difficult.

· · · · ·

SIDEBAR 7: DUTCH HARBOR POLICE BLOTTER

*Although a sense of calm has returned to the pollock fishery, conditions in Dutch Harbor are still temperamental. Dutch Harbor has a reputation as a remote outpost in the Wild West. It has been called the worst neighborhood in Ballard, due to the large number of Seattle (i.e., Ballard)–based fisher-*

men there at any one time.[53] Below are excerpts from the local Dutch Harbor police blotter for one week in July 2011.[54]

*Assistance rendered, July 11* — Complainant came to Public Safety with a fellow crew member who was suffering from paranoia.

*Ambulance request, July 11* — Caller requested EMS for a crew member who may be experiencing a heart attack aboard a fishing vessel.

*Suspicious person, July 12* — Caller reported there was a guy acting crazy in the middle of Ballyhoo Road in front of the old Alaska Ship Supply.

*Assault, July 13* — Caller reported he was having a disagreement with three to five men off the F/V [fishing vessel] Ocean Alaska at the Spit Dock which escalated in one of the men pulling a knife from his belt. Several subjects were contacted by officers, all of whom had varying descriptions of what had occurred. No charges were filed.

*Animal, July 13* — Caller reported to Public Safety that several people were feeding the eagles on East Broadway, just across the street from Kelty Field. For that reason, the eagles were not moving out of the way of oncoming vehicles, impeding traffic by staying on the roadway and making it difficult for pedestrians to safely walk their dogs. Officer responded and cleared the area.

*Suspicious activity, July 13* — A person who just came from Captains Bay Road reported a suspicious "red skiff" floating in the bay. Officers responded and determined an intoxicated individual had passed out in the skiff and went adrift. A friend responded to assist in another skiff. While crossing over to a second skiff the intoxicated individual fell into the water but was eventually pulled into the skiff by his friend.

*Ambulance request, July 13* — Officer requested EMS for a man who fell into the bay while attempting to get back onto shore.

*Suspicious activity, July 13* — Caller reported hearing what seemed like a seal bomb or gunfire coming from the bay area. Officers investigated and were unable to locate persons in the area.

*Drunken disturbance, July 16, 0149 hours* — Individual called 911 and requested officer assistance at the Harbor View bar. Report that someone had *thrown a bottle at the glass window near the dance floor. The window did not break but the bottle shattered upon impact.*

*Trespass, July 16, 0209 hours — On behalf of UniSea, the drunken individual who tried to play baseball with a bottle and hit the glass window was trespassed from all UniSea hospitality locations for three months.*

*Assault, July 16, 0254 hours — Caller reported several men fighting behind the UniSea Mall. No charges were filed.*

*Trespass, July 16, 0343 hours — On behalf of UniSea, individual female was trespassed from all UniSea bunkhouses and UniSea Inn permanently.*

*Trespass, July 16, 1645 hours — On behalf of UniSea, a male individual was trespassed from both the Cape Cheerful Lounge and the Harbor View bar due to his disruptive and aggressive behavior.*

*Ambulance request, July 16 — Caller reported a passenger on PenAir flight 7303 partially cut off a portion of his/her finger while possibly maneuvering the arm rest. Patient was safely taken to the clinic for treatment.*

.　.　.　.　.

# **12** Alaska Pollock's Challenging Future

*Those who cannot remember the past are destined to repeat it.*

GEORGE SANTAYANA

## A Storm on the Fishing Grounds

On April 6, 2011, the pollock fleet was finishing up the A-season, the first half of the pollock harvest that takes place before the fish begin to spawn. Most of the fleet was fishing northwest of the Pribilof Islands. That morning a new storm appeared on the weather maps and was intensifying as it moved toward the fleet. The US Coast Guard sent an all-points bulletin warning for all ships to take shelter. At the morning briefing, the meteorologist in attendance said that a "monster storm" would hit the next day. Winds were projected at over fifty knots with waves over thirty feet high. I spotted the report and sent the weather maps to Dr. Nick Bond, a meteorologist and NOAA's Bering Sea expert. He said it looked like the strongest storm in recent memory. What a way to end the fishing season!

As the storm approached the fishing grounds, the fleet was nearing its catch limit. The decision was made to pull their nets and to run for shelter. They made it back to Dutch Harbor before the brunt of the storm hit. They were safe.

This storm hitting the pollock fleet was a nice metaphor for the storm back in Seattle over the harvest levels for pollock that year. Months before in autumn 2010, Jim Ianelli, the lead fisheries scientist for NMFS doing the stock assessment of pollock in the eastern Bering Sea, produced his assessment that would determine catch quotas for the season. He reported a dramatic upturn in the population and recommended that quotas be increased by nearly 50%. Surveys done

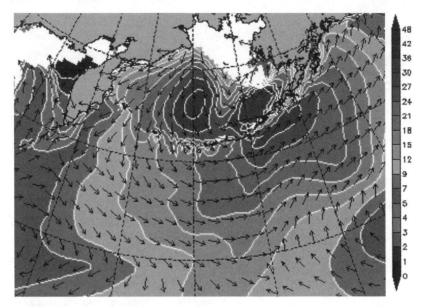

48
42
36
30
27
24
21
18
15
12
9
7
5
4
3
2
1
0

Figure 12.1. Storm of April 2011 on its way to the fishing ground. This forecast was sent from the bridge of the *American Dynasty* with the subject heading "Monster Storm" and shows wave heights projected to be over 30 feet and winds at 60 knots (flags). The US Navy, Fleet Numerical Office forecast wave heights over 50 feet.

in the summer of 2010 had found higher than expected numbers of young adults from the year class born in 2006. The surveys also indicated relatively high numbers of the 2008 year class. Just one year earlier, Ianelli's assessment of the pollock stock had reported them at their lowest level in the past thirty-two years of monitoring.

At the suggestion of a marked increase in the harvest, a hurricane from the environmental community loomed on the horizon. John Warrenchuk of Oceana said, "The pollock resource is still lower than it was in the past. They [the federal fisheries scientists] are really banking on that [2006] year class [which would join the harvest in 2011]. Those young fish are going to have to sustain the fishery for quite some time."[1] Another prominent scientist, Professor Jeremy Jackson of Scripps Institute of Oceanography, said, "The collapse of pollock would be the ultimate example of the emperor having no clothes."[2] Even some prominent fishermen questioned the amount of the harvest increases. But Ianelli seemed pretty confident about his stock projections.

Fishing for pollock in the A-season that year was reported to be strong. Another source from one of the big companies told me it was good but not great fishing. But fishing in the A-season should always be strong because the fish are schooled up in predictable locations on their way to the spawning grounds. The sophisticated acoustic instruments used to find the fish guarantee that the fishermen can find these high concentrations. However, the strong fishing was taken as evidence that the 2006 year class had produced and it seemed that federal scientist Ianelli had escaped the bullet once again.

If the fish are in short supply, the summer B-season, when fish are more dispersed and feeding, should be revealing. The progress of the 2011 B-season, the summer/autumn fishery, didn't go quite as smoothly. The bycatch of chum salmon was large, and catches started to decline in the later months, opening the question again of whether the fish stock was depleted. Or maybe they had just moved? Then late in the season, as the trawlers were trying to complete their quotas, the Chinook salmon bycatch became a problem and some vessels couldn't finish out the season. I talked with several pollock boat captains who were concerned about the stock level and the increased quotas. On August 29, the Unalaska Fishery Update reported, "The pollock B-season continues slow at a slow pace [sic], which is starting to concern a lot of folks in the community. I've been getting a lot of calls, and I'm telling them no need to panic at this time . . . very large fisheries go through slow periods at times, and the harvesters are trying their best to avoid chum salmon bycatch."[3]

Others speculated that the fish had moved north and extended over and beyond the American-Russian boundary. Over in the Russian EEZ, the fishery was harvesting young pollock right up against the border. Many fishermen and scientists were concerned that the Russians were catching the young fish that will later recruit into the US fishery.

Dr. Jim Ianelli is a razor-thin man with close-cropped hair. He sometimes sports a grizzly stubble beard. He rides an oversized mountain bike with knobby tires that announce his presence with a "whrrr" as he speeds by. He has a deep history in the ocean and in fisheries. He used to deliver sailboats from Florida to Europe, and once brought one through the Suez Canal and across the Indian Ocean to Singapore. He got a job tuna fishing in the South Pacific and later signed on with an international agency tagging tuna for research. He notes that he once held the record for tagging more tuna than anyone. These experiences led him to his master's degree in fisheries science from Humboldt State University and his PhD from the University of Washington.

Ianelli has been in charge of stock assessments of Bering Sea pollock for the AFSC since 1997. He says that his job is stressful, and I believe him. "The worst part is the uncertainty." What if things are really worse than his models tell him? That causes him to lose sleep at night. He is under intense scrutiny and pressure from the various factions. It is not a job for the faint-hearted. But Ianelli comes across as knowledgeable and confident when he talks, and carries an air of certainty. One thing that turns off the fishing industry is uncertainty. I think this has to do with decisions they themselves have to make. Fishing boat captains have to make "yes or no" judgments every day. We move or we stay? Set the trawl here or not? Run for shelter from the storm, or tough it out? Decisions are made in an instant based on experience and intuition. Wavering has no place in their world. There is no time to sit for a spell and think about it. Scientists are of a different mind. They dwell on uncertainty. Because of his confidence and open communication, Ianelli has the overwhelming respect of the fishing industry. Several people remarked that the other stock assessment scientists don't get the same respect. In Ianelli's case it has to do with knowing his numbers exceedingly well along with an outward expression of confidence, although he might not always feel that way.

One of his problems is that in the stock assessments he is required to project out two years to forecast the strength of incoming year classes. Sometimes the surveys and fishery don't see big year classes as they get close to recruiting. He cites the 1992 year class as an example. That cohort probably spent its early years in the northern Bering Sea outside of the region normally surveyed. It wasn't seen until it entered the fishery in 1996. A worse situation is a year class that looks to be strong from early results, but then doesn't develop. The result is a forecast for an increasing trend but the actual stock goes the other way, a response which could lead to disaster.

Another problem is that sometimes the two types of surveys that are conducted don't agree on the status of the stock. There is a trawl survey that samples from the bottom to 10–16 meters in the water above. Then there is an acoustic survey that samples the midwater

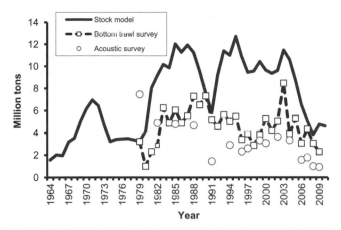

Figure 12.2. Comparison of stock assessment methodologies in the eastern Bering Sea. The stock model is for age three and older pollock. Note the general lack of agreement except the decrease in the stock from 2003 onward.

layer but misses the fish near the bottom. There is an area near bottom that the acoustic survey doesn't see, and then an area of overlap of the zone sampled by both gears. Do you add the survey results? In the area of overlap there can be double counting.[4] Or are the results of one survey more reliable than the other?

There isn't a perfect estimate of fishable biomass for the upcoming season. The estimate of age 3+ biomass for the coming year of fishing comes from the stock assessment model. But this includes an estimate of two incoming year classes that are associated with a lot of uncertainty. Then there is an estimate of spawning biomass, which doesn't include the incoming year classes. The result is uncertainty of the number in between.

The pollock stock has gone through periods of decline during Ianelli's reign as principal stock assessment scientist, but has avoided a collapse for over 40 years of fishing. Such resilience is unusual in the world of fisheries. Ianelli attributes part of his success as a fisheries stock assessment biologist to the special characteristics of pollock in the Bering Sea. For some unknown reason, in this region of the ocean the stock of pollock seem to be resilient and they have great production capacity. But Ianelli also believes that the stock needs to be relatively high for good fishing conditions.

After the hair-raising 2011 seasons, Ianelli reported that for the

2012 fishing season he doesn't think there is a declining stock, but he is uncertain of the magnitude of any increase. It turned out that the 2006 year class wasn't strong after all, only about average. For the 2012 fishing season, the recruitment of the 2008 year class is critical. Because of conflicting sources of information, Ianelli wanted to make a conservative assessment and set the incoming 2008 year class as average, rather than above average. The SSC plan team and the NPFMC did not see a reason to be conservative and recommended that the catch levels be increased above his recommendation. Overriding Ianelli's advice was a highly unusual maneuver. It was pointed out to Ianelli that his uncertainty about the catch levels would cost the industry a lot of money.

Representatives of Trident Seafood asked for a conservative approach to managing the pollock fishery in the 2012 season. Sixty captains of pollock boats were reported to have said that the 2011 quota was too high and the planned quota for 2012 also was not justified by the number of fish.[5] Many skippers reported that they couldn't find enough fish to fill their quotas in the 2011 B-season. The At-Sea Processors Association supported the higher catch rates. According to them, the low pollock catches were caused by closures due to salmon bycatch. Other industry representatives, such as the attorney-president of Glacier Fish Co., said that the problem in 2011 was that there was too much food and the fish weren't schooling up. Ianelli, on the basis of many fishermen's concerns, reduced his estimate of the size of the 2008 year class, and recommended an ABC of 1.088 million tons. The SSC decided to keep classifying the 2008 year class as "above average" and boosted the ABC to 1.22 million tons. Although that may not seem like a big difference, it is an ex-vessel (average price paid to the fishermen) difference of $120 million in wholesale value.[6] Coincidentally, several members of the SSC received grants from the at-sea processors through the Pollock Conservation Cooperative Research Center in 2012.

Some participants in the debate rumored that the companies in the industry that wanted lower quotas, such as Trident, did so because they wanted to maintain high prices for their holdings of frozen fish in Russia. Trident representatives discounted this rumor.[7]

## Shifting Winds

When you are on the high seas without another ship in sight, the ocean stretches out toward the horizon until the sky drops a veil over it. You imagine all the fish swimming beneath and it's easy to understand why a fisherman might scoff at the idea the resource is limited. How could one boat catch all those fish? That is the old view.

Nowadays, a "good fisherman" is being reshaped from the rugged risk taker to a cooperative participant. Many modern fishermen are aware that the resource is limited. It has become limited by the many vessels searching for concentrations of fish, and their improved ability to find them from a long distance. When the fish are concentrated, they can be caught efficiently, and depleted. The new fisherman is not just a harvester, but a stakeholder who needs to be aware of the conservation and science issues. The old mold might be hard to break.

In April 2011, I talked with a progressive fisherman who evolved along with the Alaska pollock fishery. Tim Thomas is the captain of American Seafoods' *Northern Jaeger*. I met Thomas in a Starbucks café at 10 a.m. When we were finished talking I looked at the clock and it read 1:48 p.m. Somehow we got focused in a zone of memories, science, and fisheries for almost four hours.

Thomas grew up on Cape Cod. He started digging clams and selling them from his front yard when he was thirteen years old. As a youth he wanted to fish, but his mother wouldn't let him. He appeased her by painting his skiff canary yellow so his mom could see him from shore, and named it "Mom's Worry."[8] He had a boat before he could drive a car. He loved the ocean and photography. After graduating from high school, he almost went to art school to be a photographer, but went fishing instead. Thomas says wistfully, "Back in those days there was a sense of romance and adventure about making your living as a fisherman. Fishing wasn't frowned upon and fishermen got a lot of respect." From his tone one gets the feeling that as the skipper of a modern factory trawler, he isn't feeling the love these days.

In 2000, Thomas won the fishing industry's most prestigious honor, the *National Fisherman*'s Highliner of the Year Award for his "overarching contributions to the industry and the community." He has been the skipper and fishing master of the *Northern Jaeger*, a 337-foot factory trawler, since it came out of the shipyards in 1989. Thomas spends a lot of time at sea. This year he'll be out on the *Jaeger* for about 220 days. He has a reputation for finding fish when no one else can.[9]

There is no mistaking that Thomas is in charge of the 110 crew members on his boat. He is lanky with reddish brown hair and big steel-blue eyes that observe the surroundings with the steady gaze of a commander. He carries himself like someone used to being listened to. Thomas sports a Spanish goatee and a simple gold wire ring in his left earlobe. The earring is so thin you have to look hard again to make sure you saw it right. He has a raspy voice with a strong twist of Cape Cod accent, and a hearty laugh which surfaces frequently and draws attention from the room.

Thomas recalled one of his strongest memories: making the move from the inshore clam fishery to offshore scalloping as a teenager. He worked on local Portuguese boats that were often pretty old and beat up. The smell of diesel and bilge permeated everything. He especially remembered that butter in the refrigerator picked up the taste of diesel. He had never gotten seasick in his small inshore boat, but in the offshore swell and diesel fumes of the scalloper he got seasick badly, and he figured he wasn't going to make it as a fisherman. Thirty years later, he fishes some of the roughest waters in the world. The roll of the boat on a swell becomes second nature to a fisherman.

In the late 1970s, Thomas was reading the *National Fisherman* and saw that a lot of nice boats were being built on the West Coast. He thought something good must be happening out there. There was opportunity. Later in the scallop fishing season, when a couple of the old local wooden boats got overloaded and flipped over in bad weather, that clinched it. He moved to the Pacific Northwest in 1978, hoping to land a job on one of the lucrative new crab fishing boats out of Anacortes. He ended up crewing for a Yugoslavian fisherman named Boris Olich, dragging for groundfish off the coast of Washington and in Puget Sound. Olich mentored him in the skill and business of fishing.

Thomas landed a job on a crabber in the Bering Sea in 1979, but the timing was bad. The king crab population was crashing. Crab fishermen were scrambling for other ways to make money. He was lucky to find a position on a trawler fishing in the pollock joint venture fishery of Shelikof Strait. He recounted a detail about how they could walk out on the codend full of pollock floating in the ocean swell behind the stern ramp because in Shelikof the pollock lived in deep water.

When they got dragged up from the bottom their swim bladders expanded from the release of pressure causing them to float. The ship's dog, Grizz, would even run out on the codend and bark at sea lions. In the Bering Sea the pollock don't float like that because they get hauled up from shallower depths and the bladders don't expand as much.

In 1980 Thomas hurt his back, and while he was recuperating he went to school and got his captain's license. When he headed back up to the Bering Sea in 1981, he fished with John Sjong on the *Arctic Trawler*. One begins to realize that there is a vast network of people in the industry connected to Røkke, Sjong, and Pereyra, the Godfathers of Ballard, and to the *Arctic Trawler*. That summer the cod fishing soured and they turned to pollock. Going from cod to pollock was a hard transition because the fillet machines weren't set up for the shape and depth of pollock. Besides, as we heard from Sjong, it was tough to market the unfamiliar fish.

Later Thomas hooked up with Einar Pedersen, another legendary fisher, and after a conversation and a handshake, he was running Einar's *Mark 1*, which worked in the joint venture fishery for Wally Pereyra's company ProFish.[10] Thomas recounted that a lot of crabbers were bankrupted or were being converted from crabbers to trawlers. He said the joke of the season was, "If you went into a bank in Ballard to open an account, the teller asked if you wanted either a toaster or a crab boat." Some months later I tried to tell that joke as a newcomer in a gathering of skippers in a boatyard and they all chimed in at the punch line. It was an old and tired joke among them.

Ocean Trawl hired Thomas to skipper the *Jaeger* in the pollock fishery. By 1996, Kjell Inge Røkke was ready to acquire more boats, and he was doing so at a staggering pace. Røkke was a pretty audacious character and it's said that he often didn't follow the conventional rules. The fishery was overcapitalized, and Røkke was able to take over Ocean Trawl's fleet of the "bird boats." The new owner, Røkke's American Seafoods, made a lot of personnel changes to the *Eagle* and *Hawk*, but the *Jaeger* consistently outperformed the other ships and her crew was kept intact.

Thomas has now skippered the *Northern Jaeger* for over twenty years. He says that he is proud of the improvements they've made in fishing in an environmentally sustainable manner. He gets animated when he talks of the bad rap that factory trawlers have gotten. He thinks that the public perception of them as Goliaths has been slanted by unfair and biased media attention. According to Thomas, their by-catch of other species is less than 1% of their total catch. The nets fish in midwater and don't hit bottom and destroy habitat, and over the years they have significantly increased their yield. The meat is made into surimi, fillets, or minced meat. The skeleton is processed into bone meal, and the leftover scraps are turned into fish meal. Fish skin is still the only part of the fish that isn't used for some product. Fish oil production from the dried and pressed heads and livers has increased by 50% with the installation of new equipment. Fresh omega-3 oil from pollock is in high demand and the *Jaeger* produces 50,000 gallons each trip.

Thomas refers to the Bering Sea pollock fishery as "my fishery" and has supported many conservation measures. In late 2010 when NMFS raised the pollock quota for the upcoming 2011 season by 50%, he stood up and advised a much smaller increase. As it turned out, the higher quota worked out fine because there were more fish present on the grounds than in the previous years. Even so, he recognized that some-thing was different about them; they were aggregating to spawn in smaller groups, in different places, and later in the season compared with previous years.

An advantage Thomas had was that he worked for and learned from some of the pioneers of the American pollock industry. He is proud of the environmental improvements in the industry that he's been a part of. Fishing is much more efficient, catches and bycatch are closely monitored, and product yield is improved. Thomas reflects that he's been lucky, he got into the fishery early, and he had good mentors.

He's made a good living from the sea, but Thomas says that things have changed for the independent fisherman. Now the fishery is closed and opportunities for the new generation are much more lim-ited. It's not as exciting anymore. There is still good money to be made by newcomers working as crew in the fishery, but nobody is going to get rich like the old days. Because of the catch shares everyone knows how much each boat can catch, so the competitive spirit is gone. The fishery is an organized and controlled business, run by big companies. The only unknown variables are the weather and the price of pollock on the market. The gold rush has ended. All the claims in Klondike have been staked and the big companies now own the fishery.

## Challenges in the Pollock Fishery

The pollock fishery developed with explosive force. Knowledge of the resource lagged far behind the process of exploitation, depletion, and discovery of new stocks. Models to manage the fish were adapted from other fishery systems before we knew almost anything about the biology of pollock. Several of the stocks declined markedly or even collapsed under fishing pressure and have yet to recover. The major stocks in the Sea of Okhotsk and eastern Bering Sea seem to be maintaining, but there is a nervous anticipation each new year as the assessments are made and the quotas set for the next year.

The pollock fishery is tightly managed now and requires infrastructure that is highly developed and expensive, a luxury enabled by the value of the fishery. There are seasonal closures to spread the fishing out over time. Area closures are in place to protect habitat, endangered species, and bycatch. Area restrictions preserve a part of the catch for specific sizes of boats. Bycatch of Chinook and chum salmon are monitored on a daily basis for hotspot closures. Efforts are underway to distinguish not only the stock of origin of pollock in the catches, but the stock of origin of the salmon bycatch. There is an industry-paid observer on every single fishing vessel and two observers on the catcher processors. Fishing positions are monitored with GPS and every catch is weighed. Information on catches is electronically submitted to NMFS on a weekly basis. Some would argue that it is too tightly managed.

But tight management means different things to different people. Some people think that "tightly managed" means closely regulated. To others it means extracting as much of the resource as possible to maximize yields and profits. In terms of sustaining the ecosystem and the resource itself, this is cutting close to the bone without understanding the consequences.

Over time, the mathematical models used to manage the pollock fishery have become very elaborate. The current model used to manage pollock in the eastern Bering Sea unabashedly estimates 772 parameters internally "conditioned on data and model assumptions."[11] That's a lot of estimating with a lot of assumptions and uncertainty. I've heard the words "over-parameterized" several times in describing these models.

As sophisticated as the new models are, there are no terms for the environment in the model, the space the fish occupy, or the animals with which they interact. The estimation of ABC doesn't consider age structure and stability, population structure, or how the population replenishes itself. These factors could be added to models. But they are also estimates with their own degrees of uncertainty. And when models get bigger with more and more parameters and assumptions, they tend to become rhapsodic and take on a life of their own. We have made improvements in what is known about the biology of pollock, but there are still problems in how to implement that knowledge in management.

Many of our current problems of managing fisheries, about which we hear so often, have a lot to do with not understanding nature and the natural history of the animals we are trying to manage. A senior stock assessment specialist at NMFS told me that "a fish is a fish, it doesn't matter if it's an anchovy or a pollock." If I seem critical, I am in good company. The well-seasoned fisheries scientist Ray Beverton remarked about how fisheries management went wrong: "Biology became subservient to the maths, in both staffing and philosophy."[12] In his book *Mismanagement of Marine Fisheries*, the eminent marine scientist Alan Longhurst said that we are overreliant on computer programs and mathematical juggling to manage our fish stocks, and we are largely ignorant of the biology and ecology that we need to manage them well. The details do matter.

### The Future of the Fishery

The emerging demand for seafood in China and India is foreboding. As these economies strengthen and individual incomes increase, people can afford to eat more seafood. US exports of frozen pollock to China have increased dramatically, from 5.3 thousand tons in 2008 to 15.4 thousand tons in 2009.[13] At current population levels, if China consumed the same amount of seafood as Japan per capita, they would consume 86 million tons (live weight) of fish, roughly equivalent to the world natural harvest. If India consumed the same amount as Americans per year, they would consume an additional 25 million tons. The seafood industry should be rubbing its hands in anticipation. Shortages in supply can drive up prices. Under a rationalized fishery, effort can be reduced. Higher prices and lower cost—for a businessman, what's not to like?

The demand for seafood already exceeds the capacity of the oceans to produce it. Meanwhile, the human population continues to increase. Technological developments have enabled aquaculture ventures to outcompete harvest fisheries in terms of cost and availability. A strategy of capture and fattening in net pens is already prevalent in the tuna industry, a prospect similar to cattle feedlots. Other fisheries can't be far behind. One envisions a brave new world of fisheries where the ocean is no longer wild, but is rather domesticated and engineered. But natural ecosystems are complex systems with poorly understood interactions and feedbacks. Most times when man tries to engineer nature, there are unintended consequences.

Political muscle has started enclosing the sea. For the independent fisherman, the writing is on the wall. Fishing for pollock is expensive. A lot of investment is required. Buying quota is costly. The result of rationalization is an increasing percentage of the profits in the hands of a few companies. This enables them to gain political power through the way our political system works. More profits means more funding for political support. The process of buying down the ships and reducing ownership concentrates that power.

One can make a good argument that the rationalization of pollock wasn't done for the sake of the fish stocks, it was done for profitability of the industry. The health of the pollock stock in the eastern Bering Sea under US management was never the issue as they were managed by a harvest limit. The bycatch and waste were manageable. But, the alternative to rationalization was to have a lot of ships and a lot of fishermen, whereby the profit is sifted into a lot of different hands, including the larger community needed to support more ships and workers. However, the concentration of profit allows more funding for political and even "scientific" influence, a positive feedback system that ensures more of the same. Some US companies are investing in processing plants in China to reduce costs and make larger profits, to the detriment of the jobs in the United States that they so convincingly argued for in passing the American Fisheries Act.

If a system better than the present pollock catch share program is devised, it would be nearly impossible to implement. The present structure of catch shares is statutory, meaning Congress would have to overhaul the law. Given a muscular industry and Congress being the way it is, one would have to ask, how likely is that? There is increasing consolidation by the larger fishing companies, their associated companies, and CDQ organizations. Not too many independent fishermen have tens of millions of dollars to spend in order to compete. Increasing profits lead to ever increasing political muscle and influence.

As ocean-harvested whitefish become even more valuable, it is predictable that there will be increasing pressure by companies to mine even more value out of the stocks. I believe that there will be increasing pressure to fish less conservatively and closer to maximum sustainable limits for all species in the Bering Sea ecosystem. The familiar call for more jobs and feeding a hungry world will be recycled, but I doubt that much pollock ends up feeding starving people in Africa or that there will be a net gain in US jobs in the fishery.

For an ecologist working in marine fisheries, the prospects are daunting. With a changing climate we are entering unknown territory. Increasing harvesting pressure on fish populations is not a cautious approach under these conditions. And if there is a collapse, we still don't have the ability to determine with certainty whether the changes are due to environmental conditions or fishing pressure.

## Conclusions

The pollock fishery has metaphorically lived a life, from its birth not too many years ago until it reached full development and maturity. In between, it experienced explosive growth and turbulent adolescence. Now it is mature, and similar to other fisheries, headed into old age.

The pollock fishery grew within the constraints and opportunities of the larger global landscape. There was the major expansion of worldwide distant water fisheries promoted by the US government. A buoyant attitude that fisheries could solve world hunger prevailed. Development of new technology enabled fishermen to find and catch their prey more efficiently. Improvements in processing and preserving fish, particularly surimi from pollock, primed a market to develop for industrialized fish products.

Harvesting pollock in the Bering Sea started as an industrial venture with the foreign high seas fishery. The key was volume, which was made possible by the sheer abundance of pollock in the sea, but also by the aggregating behavior of the fish. Eventually the political landscape changed, and the fishery was Americanized at the expense of the foreign fisheries. Initially the American fishery was dominated by highly successful individual fishermen. Big investments were needed in shoreside processing plants and funding was provided largely by Japan. Norwegian banks promoted offshore processing. The factory trawlers got bigger, more efficient, and more numerous to outcompete other fishing vessels. A political fight over the fishery ensued.

Congress took action to solve the issues of foreign ownership, overcapitalization, and competition. They legislated a catch share system. Vessels were bought out, and consolidation of the industry began. The fishery became enclosed to protect the profits of large companies. Equilibrium between the onshore and offshore components of the fishery has been reached through a political process.

The independent fishermen I interviewed were the protagonists of their own unique stories. These individuals seized an opportunity and built the American pollock fishery. Many of them were successful; some of those got bigger and incorporated. Several of those original operations are now large corporations run by businessmen. The politicalization and consolidation of the fishery has made it difficult for the small independent fishermen to survive. They have been swallowed by larger companies, or have been forced to bind together in rafts of cooperatives to have the political muscle needed to survive in a competitive environment.

At face value, the pollock story is only about a fishery. But as I see it, the fishery is an allegory of the struggle an individual faces in deciding what is right and how that changes with his circumstances. The story starts with a fisherman and what motivates him to be successful in life. He wants to be the best, to catch the most fish. After gaining success, he often adopts a different view of the world. This new view is not about being the best but about getting more. He incorporates with others to grow and compete, and becomes part of a larger organizational structure. Within a corporate organization, individual behavior changes, and the objective shifts to the needs of the company. Now the goal is to protect and gain market share. Actions are justified to benefit the corporation. A struggle ensues to find balance between what is good for the community and what is good for the business, like a tenor note searching for harmony in a chorus of bass.

*Earth provides enough to satisfy every man's need, but not every man's greed.*

MAHATMA GANDHI

As this narrative closes, the 2012 A-season has reported strong fishing once again. But as we've learned from pollock and other species, fishing in the A-season, the spawning season, can be deceptive because the fish are aggregated in predictable locations. They are easy to find and catch. The 2012 B-season would be more revealing, similar to what happened in 2011. As of this writing (October 2012), the fishing reports through summer and autumn varied from week to week, sometimes indicating relatively small catches and small fish, and other times good fishing.[14] It looks like another uncertain time for the pollock fishery.

In the past decade, the eastern Bering Sea pollock stock has declined, and the fishery has depended on strong recruitment of incoming year classes like never before. In 2011 the fishery relied on the 2006 year class for 50% of the harvest. In 2012 the fishery is projected to rely on the 2008 year class for 37% of the harvest, and that is assuming its abundance is above average, which is contentious. The diminishing 2006 year class should account for the rest of the catch.

In autumn 2012 and then again in 2013, the stock assessment of Alaska pollock will face close scrutiny when the moderate 2006 year class has run its course. During these assessments, we will get a better indication of just how strong the 2008 year class really is, and what is coming up behind it. These cohorts will need to support the fishery for several more years. We will catch a glimpse of what the future holds for Alaska pollock, and whether the seas will be calm for the pollock fishermen or if there is a storm on the horizon.

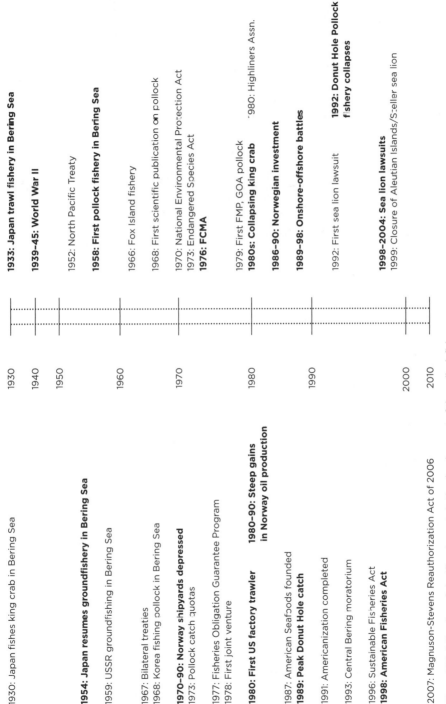

Figure 12.3. A timeline of significant events in the development of the pollock fishery.

1930: Japan fishes king crab in Bering Sea

1933: Japan trawl fishery in Bering Sea

1939–45: World War II

1952: North Pacific Treaty

1954: Japan resumes groundfishery in Bering Sea

1958: First pollock fishery in Bering Sea

1959: USSR groundfishing in Bering Sea

1966: Fox Island fishery

1967: Bilateral treaties
1968: Korea fishing pollock in Bering Sea
1968: First scientific publication on pollock

1970: National Environmental Protection Act
1970–90: Norway shipyards depressed
1973: Endangered Species Act
1973: Pollock catch quotas
1976: FCMA

1977: Fisheries Obligation Guarantee Program
1978: First joint venture

1979: First FMP, GOA pollock
1980: First US factory trawler
1980: Highliners Assn.
1980s: Collapsing king crab
1980–90: Steep gains in Norway oil production

1986–90: Norwegian investment

1987: American Seafoods founded
1989: Peak Donut Hole catch
1989–98: Onshore-offshore battles

1991: Americanization completed
1992: First sea lion lawsuit
1992: Donut Hole Pollock fishery collapses
1993: Central Bering moratorium

1996: Sustainable Fisheries Act
1998: American Fisheries Act
1998–2004: Sea lion lawsuits
1999: Closure of Aleutian Islands/Steller sea lion

2007: Magnuson-Stevens Reauthorization Act of 2006

# Appendix A: Terminology

ALLOWABLE BIOLOGICAL CATCH (ABC): as a starting point, the ABC is set equal to the fishing mortality rate at MSY leading to a harvest level (see below). It is decreased from there to account for any safety factors and uncertainties.

BOSUN: the lead fisherman or deckhand on a ship.

CATCHER PROCESSORS, OR FACTORY TRAWLERS: these are large ships from 210 to 386 feet in length that drag trawl nets behind them, near but not on the bottom. The fish are brought aboard and processed on the ship. Some ships are capable of making the meat into "surimi," a fish paste that is reprocessed into a number of products, or fillets. They also harvest the fish eggs, or roe, and make fish meal out of waste products. Other vessels specialize in filleted and frozen fish, but also make minced meat and collect roe.

CATCHER VESSELS: the smaller trawlers from 60 to 193 feet in length that deliver their catch to shore plants or to large floating processors (mother ships). Some also deliver to the catcher processors.

CODEND: the tapered and trailing end of a fishing net. Generally the mesh is finer to collect and retain fish.

COMMUNITY DEVELOPMENT QUOTA (CDQ): an amount of fish reserved for local communities to catch, sell, or lease to benefit their economic development.

FISHERY CONSERVATION AND MANAGEMENT ACT (FCMA): legislation enacted in 1976 that established the eight regional fishery management councils and the 200-mile fishery conservation zone (FCZ). In 1980 the FCMA was renamed the Magnuson Fishery Conservation and Management Act (MFCMA). The FCZ was changed to the Exclusive Economic Zone (EEZ) in 1984.

GROUNDFISH: marine fishes living on or near bottom. These include flatfishes and cods.

JOINT VENTURE (JV): a cooperative fisheries venture by which American fishing vessels would catch fish and deliver them for processing to factory mother ships of Korea, Russia, and Japan.

MAXIMUM SUSTAINABLE YIELD (MSY): the largest long-term average catch that can be taken from a stock.

NATIONAL MARINE FISHERIES SERVICE (NMFS): the agency established by Congress in 1970 to recommend fisheries harvests, establish regulations, enforce fisheries rules, and conduct research on federal fisheries resources.

THE NORTH PACIFIC FISHERY MANAGEMENT COUNCIL (NPFMC): the group established in 1976 by the Fisheries Management and Conservation and Management Act to make decisions on the harvesting of marine stocks in Alaskan waters under federal jurisdiction.

OVERFISHING: biological overfishing is to catch more fish than the stock can naturally sustain to reproduce to an expected level of abundance. Economic overfishing is to catch more fish than the fishery can sustain while maintaining profitability.

QUOTA: the share of harvest allotted to an entity.

RATIONALIZATION: the transition from a derby style of season, one with a catch quota for the whole fishery over the season leading to competition for the harvest, to a quota system whereby individual entities are assigned a percentage of the catch.

SURIMI: a gel-like paste made from minced fish meat that is reprocessed into other seafood products.

TOTAL ALLOWABLE CATCH (TAC): this is the harvest allowed by the managing agency, usually based on an "allowable biological catch" (ABC) that has been adjusted to account for ecosystem, economic, social, or other considerations.

TRAWL: the means of dragging a net through the water either on bottom (bottom trawl) or in midwater. The opening of the net is usually held open with metal "doors," wing-like devices, on either side.

# Appendix B: Other Abbreviations

AFA: American Fisheries Act

AFSC: Alaska Fisheries Science Center

ADF&G: Alaska Department of Fish and Game

BSAI: Bering Sea and Aleutian Islands

EDF: Environmental Defense Fund

EEZ: Exclusive Economic Zone

FCMA: Fishery Conservation and Management Act

FMP: Fisheries Management Plan

GOA: Gulf of Alaska

IFQ: Individual Fishing Quota

ITQ: Individual Transferable Quota

MSA: Magnuson-Stevens Act

NOAA: National Oceanic and Atmospheric Administration

QS: Quota Share

RACE: Resource Assessment and Conservation Engineering Division

SSC: Scientific and Statistical Committee

WWF: World Wildlife Fund

# Notes

PROLOGUE

1     Dear reader, here I have to admit that after nearly 40 years my memory of
      the exact dosage is misty, but I've told the story for years and sixty-four is
      the number that I have used for as long as I remember. I do distinctly recall
      that there were a lot of pills, and I counted the pills into piles of ten. One
      might ask, why so many pills instead of one goliath pill? All I can say is that
      some traditions in the Orient are mysterious.

CHAPTER ONE

1     "Centuries of Fish: Seattle's Dynamic High Seas Fisheries Fleet" (film), pro-
      duced by Bob Thorstenson (Seattle: Maritime Heritage Foundation, 2003).
2     A. M. Springer, "A Review: Walleye Pollock in the North Pacific: How Much
      Difference Do They Really Make?" Fisheries Oceanography 5 (1992): 205–23.
3     R. D. Brodeur, M. B. Decker, L. Ciannelli, et al., "Rise and Fall of Jellyfish
      in the Eastern Bering Sea in Relation to Climate Regime Shifts," Progress in
      Oceanography 77, no. 2-3 (May-Jun 2008): 103–11.
4     Ross Anderson, "Pollock Politics: A Billion-Dollar Tragedy," Seattle Times,
      August 14, 1992.
5     Hal Bernton, "Seattle Trawlers May Face New Limits on Crucial Pollock
      Fishery," Seattle Times, October 11, 2008.
6     "A Tale of Two Fisheries," Economist, September 10, 2009. http://www
      .economist.com/sciencetechnology/PrinterFriendly.cfm?story_id=14401157
      9/14/2009; and Virginia Morrell, "Can Science Keep Alaska's Bering Sea
      Pollock Fishery Healthy?" Science 326 (2009): 1340–41. http://www
      .sciencemag.org/cgi/content/full/326/5958/1340?sa_campaign=Email/
      sntw/4-December-2009/10.1126/science.326.5958.1340x.
7     Sarah Clark Stuart, Shell Game: How the Federal Government Is Hiding the
      Mismanagement of Our Nation's Fisheries (Washington, DC: Marine Fish
      Conservation Network, 2006), 1–25.
8     B. Worm et al., "Impacts of Biodiversity Loss on Ocean Ecosystem Ser-
      vices," Science 314, no. 5800 (2006): 787–90.

9    Graham Lloyd, "No-Go Scaremongers 'Fishing for Funds,'" *Australian*, March 1, 2012.

10   R. Hilborn, cited in Hal Bernton, "Will Seafood Nets Be Empty? Grim Outlook Draws Skeptics," *Seattle Times*, November 3, 2006. http://community .seattletimes.nwsource.com/archive/?date20061103&slug=

11   B. Worm, cited in Bernton, "Will Seafood Nets Be Empty?"

12   B. Worm et al., "Rebuilding Global Fisheries," *Science* 325, no. 5940 (2009): 578–85.

13   T. A. Branch et al., "Contrasting Global Trends in Marine Fishery Status Obtained from Catches and from Stock Assessments," *Conservation Biology* 25, no. 4 (Aug 2011): 777–86.

14   D. Pauly et al., "Fishing Down Marine Food Webs," *Science* 279 (1998): 860–63.

15   T. A. Branch et al., "The Trophic Fingerprint of Marine Fisheries," *Nature* 468, no. 7322 (2010): 431–35.

16   "Reports of Fisheries' Demise Are Greatly Exaggerated," At-Sea Processors Association, published 2006, www.atsea.org/learnmore.php.

17   "Norway Developing Eco-Friendly Trawl Technology," MercoPress, published April 7, 2008, http://www.mercopress.com/vernoticia.do?id=13083& formato=html.

18   Joe Haberstroh, "Tyson Goes Fishin'— After Arctic Alaska Acquisition, Company Finds Seafood, Chicken Businesses Are Oceans Apart," *Seattle Times*, May 15, 1994, http://community.seattletimes.nwsource.com/archive/ ?date19940515&slug=.

19   C. Clover, *The End of the Line* (London: Ebury Press, 2004).

20   Dave Fraser, July 9, 2011.

21   E. Bluemink, "Herring Fishermen Strike Mother Lode off Alaska," *Anchorage Daily News*, March 24, 2008.

22   J. Friedland, "Rivalries Grow for Global Fishers, as Fleets Expand and Hauls Wane," *Wall Street Journal*, interactive edition, November 25, 1997, http:// www.econ.ucsb.edu/~tedb/eep/news/fish2.ext.txt.

23   Walter Pereyra, January 9, January 12, April 11, 2011.

24   Paul Sims, "260M Servings of Supermarket Fish 'Could Be Wrongly Labelled,'" *Daily Mail*, April 25, 2011.

25   Clover, *End of the Line*.

26   Ibid.

27   A. R. Longhurst, *Mismanagement of Marine Fisheries* (New York: Cambridge University Press, 2010).

28   Clover, *End of the Line*.

29   Marine Stewardship Council, http://www.msc.org/.

30   E. L. Miles, *The US/Japan Fisheries Relationship in the Northeast Pacific: From Conflict to Cooperation?* (Seattle: Fisheries Management Foundation and Fisheries Research Institute, 1989).

31    M. Weber, *From Abundance to Scarcity* (Washington, DC: Island Press, 2001).

32    John Warrenchuk, December 17, 2010.

CHAPTER TWO

1    H. Grotius, *Mare Liberum* (The Free Sea) (Oxford: Oxford University Press, 1609). The book was originally written as a chapter in a larger volume, but was published by Elsevier as a small book independently.

2    D. Armitage, ed., *Hugo Grotius, The Free Sea, Trans. Richard Hakluyt, with William Welwod's Critique and Grotius's Reply [1609]* (Indianapolis: Liberty Fund, 2004), http://oll.libertyfund.org/title/859/66149.

3    F. T. Christy and A. Scott, *The Common Wealth in Ocean Fisheries* (Baltimore: Johns Hopkins University Press, 1965).

4    R. Milner, "Huxley's Bulldog: The Battles of E. Ray Lankester (1846–1929)," *Anatomical Record* 257, no. 3 (1999): 90–95.

5    Clark University website; http://alepho.clarku.edu/huxley/SM5/fish.html.

6    T. D. Smith, *Scaling Fisheries: The Science of Measuring the Effects of Fishing, 1855–1955* (Cambridge: Cambridge University Press, 1994).

7    Ibid.

8    M. Graham, *The Fish Gate* (London: Faber and Faber, 1949).

9    H. M. Rozwadowski, *The Sea Knows No Boundaries* (Copenhagen: International Council for the Exploration of the Sea, 2002).

10    S. Holt, "Three Lumps of Coal: Doing Fisheries Research in Lowestoft during the 1940s," talk given April 25, 2008, accessed August 3, 2012, http://www.cefas.defra.gov.uk/media/306484/sidney-holts-tallk.pdf.

11    Clover, *End of the Line*.

12    Ibid.

13    W. F. Thompson and F. H. Bell, "Biological Statistics of the Pacific Halibut Fishery, 2: Effect of Changes in Intensity upon Total Yield and Yield per Unit of Gear," *Report of the International Fisheries Commission* 8 (1934): 1–49; W. F. Thompson, *The Effects of Fishing on Stocks of Halibut in the Pacific* (Seattle: Fisheries Research Institute, University of Washington, 1950); M. D. Burkenroad, "Some Principles of Marine Fishery Biology," *Publ Inst Mar Sci Univ Tex* 2 (1951): 177–212.

14    H. Daniel and F. Minot, *The Inexhaustible Sea* (New York: Collier Books, 1961).

15    Rachel Carson, *Silent Spring* (Boston: Houghton-Mifflin, 1962); Rachel Carson, *The Sea around Us* (New York: Oxford University Press, 1950).

16    Rachel Carson, "Food from the Sea: Fish and Shellfish from New England," *Conservation Bulletin*, 1–74. Washington, DC: US Department of the Interior, US Fish and Wildlife Service, 1943.

17    Anthony Koslow, *The Silent Deep* (Chicago: University of Chicago Press, 2007).

18   J. R. Norman, *A History of Fishes* (New York: A. A. Wyn, 1948).

19   Smith, *Scaling Fisheries*.

20   Francis Bull, *Norske Portretter: Videnskapsmenn (Norwegian Portraits: Scientists)* (Copenhagen: Gyldendal, 1965).

21   V. Schwach and J. M. Hubbard, "Johan Hjort and the Birth of Fisheries Biology: The Construction and Transfer of Knowledge, Approaches, and Attitudes, Norway and Canada, 1890–1920," *Studia Atlantica* 13 (2009): 22–41.

22   Ibid.

23   Ibid.

24   Smith, *Scaling Fisheries*.

25   A. Lotka, *Elements of Mathematical Biology* (New York: Dover, 1956. Reprint of *Elements of Physical Biology*. Baltimore: Williams and Wilkins, 1925); V. Volterra, "Fluctuations in the Abundance of a Species Considered Mathematically," *Nature* 118 (1926): 558–60; A. J. Nicholson, "The Balance of Animal Populations," *Journal of Animal Ecology* 2 (1933): 131–78; H. G. Andrewartha and L. C. Birch, *The Distribution and Abundance of Animals* (Chicago: University of Chicago Press, 1954).

26   M. Graham, "Modern Theory of Exploiting a Fishery and Applications to North Sea Trawling," *Journal Conseil International pour l'Exploration Mer* 10 (1935): 264–74; M. B. Schaefer, "Some Aspects of the Dynamics of Populations Important to Management of the Commercial Marine Fisheries," *Bulletin of the Inter-American Tropical Tuna Commission* 1, no. 2 (1954): 27–56; W. E. Ricker, "Stock and Recruitment," *Journal of the Fisheries Research Board of Canada* 11 (1954): 559–623; R. J. H. Beverton and S. J. Holt, *On the Dynamics of Exploited Fish Populations: Fisheries Investigations Series II* (London: Ministry of Agriculture, Fisheries, and Food, 1957).

27   Andrewartha and Birch, *Distribution and Abundance of Animals*; Nicholson, *"Balance of Animal Populations."*

28   D. Pauly, "One Hundred Million Tonnes of Fish, and Fisheries Research," *Fisheries Research* 25 (1996): 25–38. (Tonnes is known as metric tons in the United States.)

29   Weber, *From Abundance to Scarcity*.

30   P. Molyneaux, *The Doryman's Reflection* (New York: Thunder's Mouth Press, 2005).

31   J. H. Ryther, "Photosynthesis and Fish Production in the Sea," *Science* 166 (1969): 72–76.

32   Pauly, "One Hundred Million Tonnes."

33   F. S. Russell, A. J. Southward, G. T. Boalch, and E. I. Butler, "Changes in Biological Conditions in the English Channel off Plymouth during the Last Half Century," *Nature* 234 (1971): 468–70.

34   A. W. Kendall and G. J. Duker, "The Development of Recruitment Fisheries Oceanography in the United States," *Fisheries Oceanography* 7 (1998): 69–88.

35   Weber, *From Abundance to Scarcity*.

36   Longhurst, *Mismanagement of Marine Fisheries*.

37   Ibid.

38   M. C. Finley, *All the Fish in the Sea: Fish, Fisheries Science, and Foreign Policy* (Chicago: University of Chicago Press, 2011).

39   Weber, *From Abundance to Scarcity.*

40   Ibid.

41   P. R. Josephson, *Industrialized Nature* (Washington, DC: Island Press, 2002).

42   J. W. Hubbard, "The Gospel of Efficiency and the Origins of MSY: Scientific and Social Influences on Johan Hjort and A. G. Huntsman's Contributions to Fisheries Science," Ryerson University.

43   D. L. Bottom, "To Till the Water: A History of Ideas in Fisheries Conservation," in *Pacific Salmon & Their Ecosystems: Status and Future Options*, ed. D. J. Stouder, P. A. Bisson, and R. J. Naiman (Detroit: Chapman and Hill, 1997), 569–97.

44   C. I. Zhang and S. Kim, "A Pragmatic Approach for Ecosystem-Based Fisheries Assessment and Management: A Korean Marine Ranch Ecosystem," in *Ecosystem-Based Mangement for Marine Fisheries: An Evolving Perspective*, ed. A. Belgrano and C.W. Fowler (New York: Cambridge University Press, 2011), 153–80.

45   Hubbard, "Gospel of Efficiency."

46   Clover, *End of the Line.*

47   Graham, *Fish Gate.*

48   Stratton Commission, "Our Nation and the Sea," Report of the Commission on Marine Science, Engineering, and Resources, 1969.

49   Daniel and Minot, *Inexhaustible Sea.*

50   The term *otter board* came from a type of wooden board weighted down at one end, from which fishing lines were dangled. The board was towed through the water and the angle of attachment of the towing line made it fish "out" from the shore. It is also said to have looked like an otter in the water.

51   David Butcher, *The Trawlermen* (Reading, United Kingdom: Tops'l Books, 1980).

CHAPTER THREE

1   A. Kalland, *Fishing Villages in Tokugawa Japan* (Honolulu: University of Hawaii Press, 1995).

2   A. Yatsu, "Japan," in *Impacts of Climate and Climate Change on Key Species*, ed. R.J. Beamish (Vancouver, BC: PICES Press, 2008), 57–71.

3   Chris Loew, "Surimi Import Prices Level Off in Japan," in *SeafoodSource.com*, published November 2, 2009, http://seafoodsource.com/MarketReport.aspx?id=4294976522.

4   Y. Matsuda, "History of Fisheries Science in Japan," in *Oceanographic History: The Pacific and Beyond*, ed. K. R. Benson and P. F. Rehbock (Seattle: University of Washington Press, 1993), 405–16.

5    Mathiesen and Bevan, *Soviet Fisheries*.

6    "Taiyo Fishery Company, Limited," Funding Universe, accessed August 7, 2012, http://www.fundinguniverse.com/company-histories/TAIYO -FISHERY-COMPANY-LIMITED-Company-History.html.

7    "Nippon Suisan Kaisha, Limited," Funding Universe, accessed August 14, 2012, http://www.fundinguniverse.com/company-histories/NIPPON -SUISAN-KAISHA-LIMITED-Company-History.html.

8    Matsuda, "Fisheries Science in Japan."

9    Roger Dale Smith, "Navigating from Harbored to Heavy Seas: A History of Japan's International Fisheries in the North Pacific, 1900–1976" (Master's thesis, University of British Columbia, 1999).

10   E. Miles et al., *The Management Regime of Marine Regions: The North Pacific* (Berkeley: University of California Press, 1982).

11   Finley, *All the Fish in the Sea*.

12   H. E. Gregory and K. Barnes, *North Pacific Fisheries: With Special Reference to Alaska Salmon* (San Francisco: American Council Institute of Pacific Rela- tions, 1939).

13   "Nippon Suisan Kaisha, Limited."

14   Gregory and Barnes, *North Pacific Fisheries*.

15   Ibid.

16   S. Guthrie-Shimizu, "Occupation Policy and the Japanese Fisheries Management Regime, 1945–1952," in *Democracy in Occupied Japan*, ed. M. E. Caprio and Y. Sugita (New York: Routledge, 2007), 48–66.

17   Finley, *All the Fish in the Sea*.

18   Fuchs, "Feeding the Japanese."

19   Jackson and Royce, *Ocean Forum*.

20   Fuchs, "Feeding the Japanese."

21   Guthrie-Shimizu, "Occupation Policy."

22   Ibid.

23   Jackson and Royce, *Ocean Forum*.

24   Ibid.

25   T. S. Sealy, "Soviet Fisheries: A Review," *Marine Fisheries Review* 36 (1974): 5–33.

26   A. Nishimura, email, January 15, 2011.

27   H. Kasahara and W. Burke, "North Pacific Fisheries Management," in *The Program of International Studies of Fishery Arrangements* (Washington, DC: Resources for the Future, 1973), 1–91.

28   H. Kasahara, *Fisheries Resource of the North Pacific Ocean, Part 1: A Series of Lectures Presented at the University of British Columbia, January and February, 1960* (Vancouver: University of British Columbia, 1961).

29   Nishimura; H. Uchida and M. Watanabe, "Walleye Pollack [sic] (Suketou- dara) Fishery Management in the Hiyama Region of Hokkaido, Japan," in *Case Studies in Fisheries Self-Governance*, ed. R. Townsend, R. Shotton, and H. Uchida, FAO Fisheries Technical Paper (Rome: FAO, 2008), 163–74.

30  Uchida and Watanabe, "Walleye Pollack [sic]."

31  Nishimura.

32  Kasahara and Burke, "North Pacific Fisheries Management."

33  Kasahara, *Fisheries Resource of the North Pacific Ocean.*

34  Jackson and Royce, *Ocean Forum.*

35  Mathiesen and Bevan, *Soviet Fisheries.*

36  Tarleton Bean as cited in J. W. Collins, "Report on the Fisheries of the Pacific Coast of the United States," in *Report of Commissioner for 1888* (Washington, DC: US Commission of Fish and Fisheries, 1892), 3–269.

37  Ibid.

38  Pacific Seafood Processors Association, *A Strategy of the Americanization of the Groundfish Fisheries of the Northeast Pacific: Summary Report* (Seattle: Pacific Seafood Processors Association, 1985).

39  Kasahara, *Fisheries Resource of the North Pacific Ocean.*

40  S. C. Sonu, "Surimi," *NOAA Technical Memorandum NMFS,* January 1986.

41  R. A. Fredin, "History of Regulation of Alaska Groundfish Fisheries," in *Processed Report* (Seattle: Northwest and Alaska Fisheries Center, 1987), 63.

42  Ibid.

43  Clem Tillion interview in "Centuries of Fish."

CHAPTER FOUR

1  Ann Touza, "Fishing on the Far Side of the World," *Pacific Fishing,* April 2011, 9–12.

2  I was never able to make contact with Dave. Many people said he sold everything and moved to the desert in California. This story is pieced together from the book *Lost at Sea* by Patrick Dillon (New York: Touchstone, 1998), and interviews with several people in the industry.

3  Tor Tollesen, March 23 and April 23, 2011.

4  The *Aleutian Speedwell* was obtained by Trust Company of the West in 1996. It was managed by International Maritime Management, run by Harald Holmen, a former credit officer of Christiana Bank (in 1994), who held the original mortgage. In 1997 it was acquired by American Seafoods and renamed the *Christina Ann.* The ship was prohibited from further fishing in US waters after passage of the American Fisheries Act.

5  Wesley Loy, "War Horse: Clem Tillion Scores Another Victory at Adak," *Pacific Fishing,* August 2004, 31–32.

6  "Centuries of Fish."

7  E. Miles, April 14, 2011.

8  Weber, *From Abundance to Scarcity.*

9  Most of the following discussion was taken from interviews with Walter Pereyra, January 9, January 12, April 11, 2011.

10  R. C. Toth, "Attack May Have Caught Kremlin by Surprise," *Seattle Times,* September 2, 1983.

11   With the downfall of the Soviet Union, Vneshtorgbank became Bank VTB, one of the largest in Russia.

12   Pereyra; Mick Stevens, February 21, 2011.

13   D. L. Alverson, September 12, 2007.

14   Dave Fraser, July 9, 2011.

15   Sheila Shafer, "Barry Fisher: Still Outspoken after All These Years," *Pacific Fishing*, March 1986, 31–39.

16   Ibid.

17   Fraser.

18   Shafer, "Barry Fisher."

19   R. Hornnes, "Norwegian Investments in the US Factory Trawler Fleet, 1980–2000" (Master's thesis, University of Bergen, 2006).

20   "Centuries of Fish."

21   D. L. Alverson, *Race to the Sea* (New York: iUniverse, 2008).

22   Ibid.

23   D. L. Alverson, "Technical Report 1: The Pacific Northwest Bottomfish Resources and Fisheries—a Historical Perspective," in *A Strategy for the Americanization of the Groundfish Fisheries of the Northeast Pacific*, Saltonstall-Kennedy Grant Program (Seattle: Pacific Seafood Processors Association,1985).

24   Pacific Seafood Processors Association, *Americanization of Groundfish Fisheries*.

25   Ibid.

26   "US Surimi Plant in Production," *Fishing News International*, June 1986, 1.

27   Susan Jelley, "The Price of Fishing in American Waters," *Fishing News International*, May 1983, 8–9.

28   "Surimi Giant for US," *Fishing News International*, September 1986, 1.

29   Hornnes, "Norwegian Investments."

30   NOAA Fisheries, "Final Environmental Impact Statement for American Fisheries Act, Amendments 61/61/13/8," published 2002, http://www.fakr .noaa.gov/sustainablefisheries/afa/eis2002.pdf.

31   Hornnes, "Norwegian Investments."

32   "Centuries of Fish."

CHAPTER FIVE

1   C. Mullon, P. Freon, and P. Cury, "The Dynamics of Collapse in World Fisheries," *Fish and Fisheries* 6 (2005): 111–20.

2   Smith, Scaling Fisheries.

3   T. Mangelsdorf, *History of Steinbeck's Cannery Row* (Santa Cruz: Western Tanager Press, 1986).

4   A. F. McEvoy, *The Fisherman's Problem* (Cambridge: Cambridge University Press, 1986).

5    M. H. Glantz, "Science, Politics, and Economics of the Peruvian Anchoveta Fishery," *Marine Policy* 3 (1979): 201–10.

6    W. E. Schrank, "The Newfoundland Fishery: Ten Years after the Moratorium," *Marine Policy* 29 (2005): 407–20.

7    Pauly et al., "Fishing Down Marine Food Webs"; and K. T. Frank, B. Petri, and N. L. Shackell, "The Ups and Downs of Trophic Control in Continental Shelf Ecosystems," Trends in Ecology and Evolution 22 (2007): 236–42; and Springer, "How Much Difference?"

8    Springer, "How Much Difference?"; and R. L. Merrick, T. R. Loughlin, and D. G. Calkins, "Decline in Abundance of the Northern Sea Lion, Eumetopias Jubatus, in Alaska," Fishery Bulletin 85 (1987): 351–65.

9    The estimate may also contain some unknown portion of the shelf stock that may have "wandered" into the Basin and been caught there (J. Ianelli, personal communication). Another index estimates a much lower maximum biomass of pollock in the Aleutian Basin, about 4 million tons (CBS 2010); this index takes the abundance of fish determined by an acoustic survey near Bogoslof Island during a two-week period in February-March and assumes that they represent 60% of the total Aleutian Basin population. My view is that this method is less reliable because the area surveyed in February-March is small relative to the overall distribution of pollock at that time (K. Okada, "Biological Characteristics and Abundance of the Pelagic Pollock in the Aleutian Basin," paper presented at the International North Pacific Groundfish Symposium [Japan: Far Seas Fisheries Research Laboratory, 1983]; also T. J. Mulligan, K. M. Bailey, and S. Hinckley, "The Occurrence of Larval and Juvenile Walleye Pollock, *Theragra Chalcogramma*, in the Eastern Bering Sea with Implications for Stock Structure," *Proc. Int. Symp. Biol. Mgmt. Walleye Pollock, Alaska Sea Grant Report* [1989]: 471–90; also S. Hinckley, "Spawning Dynamics and Fecundity of Walleye Pollock [*Theragra chalcogramma*] in the Eastern Bering Sea" [Master's thesis, University of Washington, 1986]). I also think that this method is less reliable because catches in the Bogoslof Island region were very small relative to those in the rest of the Basin, and the 60% assumption has little basis for assignment. Furthermore, from one of the few pre-exploitation surveys of the Basin, Okada reported a biomass of pollock in the Aleutian Basin of up to 5.4 million tons in 1978, prior to the reported increase in abundance in the 1980s (Okada 1979; cited in Okada paper of 1983).

10   K. R. Smedbol and J. S. Wroblewski, "Metapopulation Theory and Northern Cod Population Structure: Interdependency of Subpopulations in Recovery of a Groundfish Population," *Fisheries Research* 55 (2002): 161–74.

11   Okada, "Biological Characteristics and Abundance."

12   L. Fritz, "Trawl Locations of Walleye Pollock and Atka Mackerel Fisheries in the Bering Sea, Aleutian Islands, and Gulf of Alaska from 1977–92," in *AFSC Processed Report* 93–08 (Seattle: Alaska Fisheries Science Center, 1993).

13    Joel Gay, "Filling the Doughnut Hole," *Pacific Fishing*, November 1992, 24–25.

14    D. Parker, "US and Soviets Seek Plug for Donut Hole," *Pacific Fishing*, October 1991, 37.

15    CBS 2010.

16    T. Sasaki and T. Yoshimura, "Past Progress and Present Condition of the Japanese Pollock Fishery in the Aleutian Basin," document submitted to the Annual Meeting of the International North Pacific Fisheries Commission, Vancouver, Canada, October 1987 (Tokyo: Fisheries Agency of Japan, 1987).

17    N. Akira et al., "Interannual Variability in Growth of Walleye Pollock, *Theragra Chalcogramma*, in the Central Bering Sea," *Fisheries Oceanography* 10 (2001): 367–75. Figure 7 in this article shows that the temperature was varying but there was no trend during the period of question.

18    G. L. Hunt and K. F. Drinkwater, "Background on the Climatology, Physical Oceanography, and Ecosytems of the Sub-Arctic Seas: Appendix to the Essas Science Plan," *GLOBEC Report* 96 (2005). Figure 49 in this report shows that zooplankton biomass was varying but there was no trend over the period in question.

19    Alverson, *Race to the Sea*.

20    Finley, *All the Fish in the Sea*.

21    NOAA Fisheries, "Final Environmental Impact Statement."

22    The first value is from "The Doughnut-Hole Pact—Overfished Bering Sea Will Get a Welcome Break," *Seattle Times*, August 27, 1992, http://community.seattletimes.nwsource.com/archive/?date=19920827&slug=1509661; the second value is from C. Miller, "How Many Pollock in Donut Hole?" Alaska Fisherman's Journal, October 1987, 54–56.

23    P. Greenberg, *Four Fish: The Future of the Last Wild Food* (New York: Penguin Press, 2010).

24    Miles, "US/Japan Fisheries Relationship."

25    "Hole Used for Illegal Fishing," *Fishing News International*, June 1988, 31.

26    Miles.

27    Gay, "Filling the Doughnut Hole." It isn't explained how US officials knew what unobserved boats were catching.

28    "Talks Heat Up on Illegal Japanese Fishing," *Pacific Fishing*, November 1988, 19.

29    Ibid.

30    A. Vaisman, "Trawling in the Mist: Industrial Fisheries in the Russian Part of the Bering Sea," TRAFFIC Network Report. Published 2001. http://www.traffic.org/fisheries-reports/traffic_pub_fisheries5.pdf.

31    R. Baird, "Illegal, Unreported, and Unregulated Fishing: An Analysis of the Legal, Economic, and Historical Factors Relevant to Its Development and Persistence," *Melbourne Journal of International Law* 5 (2004): 299–334, http://www.austlii.edu.au/au/journals/MelbJIL/2004/13.html.

32    Miles.

33    Ross Anderson, "The Politics of Pollock," *Seattle Times*, February 12, 1991.

34    Alverson, Race to the Sea.

35    Miles.

36    Alverson.

37    V. G. Wespestad, "The Status of Bering Sea Pollock and the Effect of the 'Donut Hole' Fishery," *Fisheries* 18(3) 1993: 18–24.

38    J. N. Ianelli, T. Hokalehto, and N. Williamson, "An Age-Structured Assessment of Pollock (*Theragra Chalcogramma*) from the Bogoslof Island Region," *Stock Assessment and Fishery Evaluation Report* (2006): 201–36, http://www.afsc.noaa.gov/refm/docs/2006/BOGpollock.pdf.

39    J. Ianelli, personal communication, August 2010.

40    J. Hjort, "Fluctuations in the Year Classes of Important Food Fishes," *Journal du Conseil* 1 (1926): 5–38.

41    M. Storr-Paulsen et al., "Stock Structure of Atlantic Cod (*Gadus morhua*) in West Greenland Waters: Implications of Transport and Migration," *ICES Journal of Marine Science* 61 (2004): 972–82.

42    P. Fauchald, M. Mauritzen, and H. Gjosaeter, "Density-Dependent Migratory Waves in the Marine Pelagic Ecosystem," *Ecology* 87 (2006): 2915–24.

43    A. Corten, "A Proposed Mechanism for the Bohuslän Herring Periods," *ICES Journal of Marine Science* 56 (1999): 207–20.

44    D. H. Cushing, *Marine Ecology and Fisheries* (Cambridge: Cambridge University Press, 1975).

45    Okada, "Biological Characteristics and Abundance"; Hinckley, "Spawning Dynamics and Fecundity"; T.J. Mulligan, K.M. Bailey, and S. Hinckley, "Implications for Stock Structure"; and J. J. Traynor et al., "Methodology and Biological Results from Surveys of Walleye Pollock (*Theragra Chalcogramma*) in the Eastern Bering Sea and Aleutian Basin in 1988," paper presented at the Proceedings of the Symposium on Application of Stock Assessment Techniques to Gadids, Seattle, Washington, 1989.

46    D. E. Schindler et al., "Population Diversity and the Portfolio Effect in an Exploited Species," *Nature* 465 (2010): 609–12.

47    H. Svedäng, M. Cardinale, and C. Andre, "Recovery of Former Fish Productivity: Philopatric Behaviors Put Depleted Stocks in an Unforseen Deadlock," in *Ecosystem-Based Management for Marine Fisheries: An Evolving Perspective*, ed. A. Belgrano and C. W. Fowler (New York: Cambridge University Press, 2011), 232–47.

48    E. P. Ames, "Atlantic Cod Stock Structure in the Gulf of Maine," *Fisheries* 29, no. 1 (2004): 10–28.

49    P. M. Brooks and U. R. Sumaila, "Without Drastic Measures, Gulf of Maine Cod Fishery Will Be Lost Forever," *Portland Press Herald/Maine Sunday Telegram*, April 2, 2006,http://www.seaaroundus.org/newspapers/2006/MaineSundayTelegram_April2006.pdf.

50    Ames, cited in Akira et al., "Interannual Variability."

51    Corten, "Proposed Mechanism."

52    CBS 2007: 12th Annual Conference of the Parties to the Convention on the

Conservation and Management of Pollock Resources in the Central Bering Sea. Virtual conference hosted by China, http://www.afsc.noaa.gov/REFM/ CBS/Docs/12th%20Annual%20Conference/12th%20Annual%20Donut%20 Hole%20Final%20Report.pdf.

CHAPTER SIX

1    John Sjong, January 28, 2011.
2    Vera Schwach, March 6, 2011.
3    Ibid.
4    Bill Saporito, "The Most Dangerous Job in America," in CNN Money, published May 31, 1993, http://money.cnn.com/magazines/fortune/fortune _archive/1993/05/31/77905/index.htm.
5    Bill Saporito, "The Most Dangerous Job in America," in CNN Money, published May 31, 1993, http://money.cnn.com/magazines/fortune/fortune _archive/1993/05/31/77905/index.htm.
6    A. Tollefsen, *Following the Waters* (Brewster, MA: Leifur Publications, 2005).
7    The older pre–North Sea oil generation of Norwegian immigrants sometimes refer to the more recent arrivals as "oil babies."
8    "Centuries of Fish."
9    Bruce Ramsey, "The Tide of Change: Bering Sea Spawns New Fishing Era," *Seattle Post-Intelligencer*, May 2, 1988.
10   Gunnar Stavrum, *Kjell Inge Røkke: En Uautorisert Biografi* (Oslo: Glydendal, 1997).
11   Sjong.
12   Tollesen, March 23 and April 23, 2011.
13   Hornnes, "Norwegian Investments."
14   Ibid.
15   NOAA Fisheries, "Final Environmental Impact Statement."
16   Hornnes, "Norwegian Investments."
17   Ibid.
18   Uri was a successful crab fisherman who was a second-generation Norwegian-American. His parents were from Sykkylven, like Sjong. Oyvind Malmin, "Norwegian Americans in the King Crab Fishery" (Master's thesis, University of Bergen, 2008).
19   "Centuries of Fish."
20   Sjong.
21   Sverre Arestad, "Norwegians in the Pacific Coast Fisheries," *Norwegian American Studies* 30 (1985): 13.
22   "Centuries of Fish."
23   In 1995, the Arctic Trawler left the eastern Bering Sea to fish for pollock off the coast of Russia because it couldn't compete in the combat-style fishing of that area. The ship was now owned by Arctic King Fisheries, a subsidiary of Kaioh International Investment, itself a subsidiary of Kaioh

Suisan. She was flying the flag of Belize and was involved in a joint venture fishery, which was also unprofitable. The Arctic Trawler returned to the United States in 1997 but had no domestic catch history after 1995. She was tied up and didn't fish. Moorage alone cost $600,000–$900,000 per year. Then in 1998, she was bought by Trinity Seafoods for $2 million. When the American Fisheries Act passed in 1998 dividing up the pollock resource into private quotas, the Arctic Trawler, without a recent history of fishing, didn't get a share. After being left out of the pollock fishery, Arctic King and Trinity renegotiated the sale price to $750,000, and United States Seafood took title. She was renamed the Sea Freeze Alaska. She currently fishes to supply the "headed and gutted" market for Alaska groundfish. The evolving story of the old Arctic Trawler is symbolic of many of the vessels involved in the Bering Sea fishery. Try to be at the right place and the right time, and if not, then adapt to take advantage of other opportunities.

24 The *Royal Sea* now fishes for American Seafoods as the *Katie Ann*.

25 "Skipper's to Buy Alaska Pollock from Royal Sea," *Pacific Fishing*, November 1986, 21.

26 RGI and American Seafoods are somehow closely connected to Norwegian Seafoods and the Aker companies, controlled by Røkke. A corporate veil makes it difficult to determine the interconnectedness of these companies. Currently Aker Biomarine, which Røkke controls, has an office in the same building as American Seafoods. The former owner of Oceantrawl, taken over by RGI, lists his office in LinkedIn as in the same building and on the same floor as American Seafoods. Bernt Bodal, president of American Seafoods, list himself as a former president of RGI. Uncovering these relationships might be a book-length project in itself.

27 Beth McGinley, "Erik Breivik," *Pacific Fishing*, December 1988, 50–55.

28 Erik Breivik, November 14, 2011.

29 H. M. Sudness, "From Bare Hands to Global Embrace," *World Fishing* 55, no. 2 (March 2006): 10–11.

30 Torgeir Anda, *Røkke* (Oslo: Universitetsforlaget, 1997).

31 Stavrum, *Kjell Inge Røkke.*

32 Joel Gay, "Will Factory Trawlers Last into the Future?" *Pacific Fishing*, October 1992, 48–56.

33 Stavrum, *Kjell Inge Røkke.*

34 Ibid.

35 Ibid.

36 "Acona: Surimi and Fillets from Alaska to New Zealand," *Fishing News International*, July 1990, 44–49.

37 "Rise of American Seafoods Company," *Fishing News International*, October 1991, 16–21.

38 Eventually the *Golden Alaska* was rebuilt and is now owned by Golden Alaska Seafoods.

39 "Chronicle of a Plunder Foretold," published November 17, 1997, http://

archive.greenpeace.org/comms/fish/plunsum.html; and "Roekke and RGI," Greenpeace, accessed August 3, 2012, http://archive.greenpeace.org/comms/fish/am03.html.

40 Anda, *Røkke*.

41 Nina Berglund, "Labour's Capitalist Darling Pushes His Luck," News in English, published April 21, 2009, http://www.newsinenglish.no/2009/04/21/labour%E2%80%99s-capitalist-darling-pushes-his-luck/.

42 "From Bare Hands to Global Embrace," World Fishing and Aquaculture, published February 27, 2006, http://www.worldfishing.net/comment-and-analysis101/interviews/from-bare-hands-to-global-embrace.

43 Bernt Bodal, February 9, 2011.

44 Jorn Madslien, "Norway Tycoon Gets Jail for Boat Crime," in *BBC News*, published July 1, 2005, http://news.bbc.co.uk/2/hi/business/4641781.stm.

45 Andre Nilsen, "The Norwegian Parliament Awards Prize to Corrupt Guantanamo Operator," Oxford Council on Good Governance, 2006.

46 Sudness, "From Bare Hands to Global Embrace."

47 Bodal.

48 Wesley Loy, "Making Waves—How Bernt Bodal Navigated His Company to the Lead of the Pollock Fishery," *Anchorage Daily News*, September 11, 2005.

49 The island is also the origin of the copper mined to construct the Statue of Liberty. There is a small replica of the monument overlooking the bay in Visnes on Karmøy.

50 Astrid Tollefsen, *Following the Waters: Voices from the Final Norwegian Emigration* (Brewster, MA: Leifur Publications, 2005).

51 Kaare Ness, October 28, 2011.

52 Otto Von Munchow, "Norwegian American: Kaare Ness," published 2007, Norway.Com.

53 Tollesen.

54 D. L. Alverson, February 18, 2011.

55 Saporito, "Most Dangerous Job."

56 Touza, "Fishing on the Far Side."

57 Ibid.

CHAPTER SEVEN

1 P. E. Stephan and S. G. Levin, *Striking the Mother Lode in Science: The Importance of Age, Place, and Time* (New York: Oxford University Press, 1992).

2 A. D. MacCall, *Dynamic Geography of Marine Fish Populations* (Seattle: University of Washington Press, 1990).

3 G. I. Izhevskii, *Forecasting of Oceanological Conditions and the Reproduction of Commercial Fish*, trans. Israel Program for Scientific Translations (Moscow: All-Union Scientific Research Institute of Marine Fisheries and Oceanography, 1964).

4 Most of this published information can be found in the following: K. M.

Bailey et al., "Population Structure and Dynamics of Walleye Pollock, *Theragra Chalcogramma*," *Advances in Marine Biology* 37 (2000): 179–255; and K. M. Bailey et al., "Recruitment of Walleye Pollock in a Physically and Biologically Complex Ecosystem: A New Perspective," *Progress in Oceanography* 67, no. 1–2 (2005): 24–42.

5     I was the author of one of them.

6     Jon Egil Skjæraasen et al., "Extreme Spawning-Site Fidelity in Atlantic Cod," *ICES Journal of Marine Science* 68, no. 7 ( 2011 ): 1472–77.

7     S. M. Sogard and B. L. Olla, "Effects of Light, Thermoclines, and Predator Presence on Vertical Distribution and Behavioral Interactions of Juvenile Walleye Pollock, *Theragra Chalcogramma* Pallas," *Journal of Experimental Marine Biology and Ecology* 167 (1993): 179–95; also S. M. Sogard and B. L. Olla, "Food Deprivation Affects Vertical Distribution and Activity of a Marine Fish in a Thermal Gradient: Potential Energy-Conserving Mechanisms," *Marine Ecology Progress Series* 133 (1996): 43–55.

8     Brodeur et al., "Rise and Fall of Jellyfish."

9     J. A. Estes et al., "Killer Whale Predation on Sea Otters Linking Oceanic and Nearshore Ecosystems," *Science* 282 (1998): 473–76.

10    D. P. DeMaster et al., "The Sequential Megafaunal Collapse Hypothesis: Testing with Existing Data," *Progress in Oceanography* 68, no. 2–4 (2006): 329–42.

11    P. A. Balykin and A. V. Buslov, "Long-Term Variability in Length of Walleye Pollock in the Western Bering Sea and East Kamchatka," *PICES Scientific Report* 20 (2002): 67–69.

12    G. Davis, "US and Russia Team Up in Bering Sea," *Pacific Fishing*, Yearbook 1994, 28.

13    Longhurst, *Mismanagement of Marine Fisheries*.

14    "Review of the State of World Fishery Resources: Marine Fisheries. Northwest Pacific," *FAO Fisheries Circular No. 920 FIRM/C920*, 1997.

15    "Alaska Pollock–West Bering Sea," FishSource, published 2010, http://www.fishsource.org/fishery/data_summary?fishery=Alaska+pollock+-+W+Bering+Sea.

16    S. Conover and V. Monakhov, "Russia Review," *Pacific Fishing*, March 1993, 19.

17    B. N. Kotenev and O. A. Bulatov, "Dynamics of the Walleye Pollock Biomass in the Sea of Okhotsk," *PICES Scientific Report* 36 (2009): 291–95.

18    Chris Bird, "Russia's Favorite Fish on Verge of Extinction," *New Scientist* 140, no. 1902 (1993): 10.

19    "Alaska Pollock–Sea of Okhotsk," FishSource, published 2010, http://www.fishsource.com/fishery/identification?fishery=Alaska+pollock+-+Sea+of+Okhotsk+%28Country%3A+RU%3B+Gear%3A+TM%3B+MSC-Client%3A+RPA%3B+MSC-Status%3A+MSC+Full+Assessment%3B%29.

20    Vaisman, "Trawling in the Mist"; also Alyona Sokolova, "Seattle Fish Exec to Russians: 'Your Problem Is Corruption,'" *Pacific Fishing*, October 2006, 1.

21     World Wildlife Fund (WWF), "Illegal Fishers Plunder the Arctic," published April 16, 2008. http://wwf.panda.org/wwf_news/press_releases/?131061.

22     "Review of the State of World Fishery Resources."

23     "Alaska Pollock–Japanese Pacific," FishSource, published 2010, http://www .fishsource.org/fishery/identification?fishery=Alaska+pollock+-+Japanese +Pacific.

24     H. Miyake et al., "Present Condition of Walleye Pollock Spawning Ground Formation in the Sea of Japan off Western Hokkaido, Viewed from the Recent Conditon of the Egg Distributions," *Bulletin of the Japanese Society of Fisheries Oceanography* 72 (2008): 265–72.

## CHAPTER EIGHT

1     Longhurst, *Mismanagement of Marine Fisheries.*

2     W. M. Getz and R. G. Haight, *Population Harvesting* (Princeton, NJ: Princeton University Press, 1989).

3     R. C. Francis, "Fisheries Science Now and in the Future: A Personal View," *New Zealand Journal of Marine and Freshwater Research* 14 (1980): 95–100.

4     Pauly, "One Hundred Million Tonnes."

5     Weber, *From Abundance to Scarcity.*

6     Clover, *End of the Line.*

7     Longhurst, *Mismanagement of Marine Fisheries.*

8     These simple models are still the basis of management decisions. Virtually every fisheries scientist and economist I've talked to about harvest rates has started drawing dome-shaped isopleths of yields versus harvest rates as if it were scripture.

9     Longhurst, *Mismanagement of Marine Fisheries.*

10     Whether depensation occurs in natural populations or not is controversial. Often meta-analyses, which may compare many populations or aggregate populations or even different species, are used to test the concept. The meta-analysis approach is frequently used in fisheries science, which may be why there is so much controversy about the status of global fisheries and the basic ecological processes underlying fish population dynamics. In the comparative and meta-analysis approach, the variability in life histories and local environments obfuscates, rather than clarifies, the underlying relationships.

11     Bernard A. Megrey and V. G. Wespestad, "Alaska Groundfish Resources: 10 Years of Management under MGCMA," *North American Journal of Fisheries Management* 10, no. 2 (1990): 125–43.

12     Worm et al., "Impacts of Biodiversity Loss."

13     R. A. Myers and B. Worm, "Rapid Worldwide Depletion of Predatory Fish Communities," *Nature* 423 (2003): 280–83.

14     L. Watling and E. Norse, "Disturbance of the Seabed by Mobile Fishing Gear: A Comparison to Forest Clearcutting," *Conservation Biology* 12 (1998): 1180–97.

15   Worm et al., "Impacts of Biodiversity Loss."

16   Ray Hilborn, "Reinterpreting the State of Fisheries and Their Management," *Ecosystems* 10, no. 8 (2007): 1362–69.

17   Branch et al., "Contrasting Global Trends in Marine Fishery Status."

18   Hilborn, "Reinterpreting."

19   R. Hilborn, "*Mismanagement of Marine Fisheries* by Alan Longhurst" (review), *Fisheries* 36, no. 5 (2011): 247.

20   Alan Longhurst, personal communication. Email: November 1, 2011.

21   Robert McClure, "Jellyfish for Lunch? It's No Joke, Says Scientist," *Seattle Post-Intelligencer*, May 3, 2004; and Mike Urch, "Time to Be Upbeat about Fisheries," SeafoodSource.com, published November 7, 2011, http://www .seafoodsource.com/newsarticledetail.aspx?id=12841.

22   D. Pauly, "Aquacalypse Now," *New Republic*, September 28, 2009, http:// www.tnr.com/article/environment-energy/aquacalypse-now.

23   Marine Fish Conservation Network (Stuart, "Shell Game").

24   Branch et al., "Contrasting Global Trends in Marine Fishery Status."

25   Alverson; Pereyra.

26   D. D. Huppert, "Managing the Groundfish Fisheries of Alaska: History and Prospects," *Reviews in Aquatic Sciences* 4, no. 4 (1991): 339–73.

27   PEW Oceans Commission, *A Dialogue on America's Fisheries* (Arlington, Virginia: PEW Oceans Commission, 2003).

28   F. J. Mueter et al., "Expected Declines in Recruitment of Walleye Pollock (*Theragra Chalcogramma*) in the Eastern Bering Sea under Future Climate Change," *ICES Journal of Marine Science* 68, no. 6 (Jul 2011): 1284–96.

29   Ibid.

CHAPTER NINE

1   "Raising a Ruckus!" Living on Earth, published June 5, 1998, http://www .loe.org/shows/segments.html?programID=98-P13-00023&segmentID=5.

2   Sean Cavanagh, "Greenpeace Ends Bridge Protest—Weekend Effort Turns Back Fishing Trawler," *Seattle Times*, August 18, 1997.

3   "Raising a Ruckus!"

4   Danny Westneat, "Greenpeace Blockades Trawlers—Activists Chain Propellers to Halt Fishing Trips," *Seattle Times*, August 17, 1996.

5   Joseph Plesha, July 25, 2011.

6   "Senator Seeks Limits on Fleet of Huge Bering Sea Trawlers," *Seattle Times*, December 15, 1997.

7   "9 Arrested in Protest of Factory Fishing," *Seattle Times*, August 2, 1995.

8   Westneat, "Greenpeace Blockades Trawlers."

9   Ross Anderson, "Factory Trawlers Facing Rough Seas-Greenpeace, Political Foes Line Up against Big Boats," *Seattle Times*, January 9, 1998.

10   D. Helvarg, "Full Nets, Empty Seas—Harmful Effects of Supertrawlers on Ocean Fisheries," *Progressive*, November 1997.

11   Jebb Wyman, "Fracas over US-Russian Boundary," *Pacific Fishing*, November 1999, 26; D. Knight, "Super Trawler Threatens Marine Food Chain," Inter Press Service, published January 13, 1998, http://www.ips.fi/koulut/ 199803/11.htm.

12   "Megatrawl Developed for Pollock Fishery," *Fishing News International*, March 1989, 82–83.

13   Knight, "Super Trawler Threatens Marine Food Chain."

14   Ross Anderson, "Big Factory Trawlers: To Be or Not to Be? Hearings Begin Tomorrow on Possible Ban of Vessels," *Seattle Times*, March 25, 1998.

15   Knight, "Super Trawler Threatens Marine Food Chain."

16   Fred Munson, March 15, 2011.

17   P. A. Larkin, "An Epitaph for the Concept of Maximum Sustainable Yield," *Transactions of the American Fisheries Society* 106 (1977): 1–11.

18   Ken Stump, January 19, 2011.

19   J. McBeath, "Greenpeace v. National Marine Fisheries Service: Steller Sea Lions and the Commercial Fisheries in the North Pacific," *Alaska Law Review* 21 (2004): 1–42; also L. Fritz et al, "The Threatened Status of Steller Sea Lions, Eumetopias Jubatus, under the Endangered Species Act: Effects on Alaska Groundfish Fisheries Management," *Marine Fisheries Review* 57, no. 2 (1995): 14–27.

20   McBeath, "Greenpeace v. National Marine Fisheries Service."

21   Fritz et al., "Threatened Status of Steller Sea Lions."

22   "Pollock Cutbacks Sought—Environmentalists Want to Protect Sea Lions," *Seattle Times*, August 10, 1999.

23   McBeath, "Greenpeace v. National Marine Fisheries Service."

24   Ibid.

25   Ibid.

26   Associated Press, "Fishing Restrictions to Aid Sea Lions Target of Hearing," *Seattle Times*, October 17, 2011.

27   Mary Anne Hansan, "Open Letter to Journalists from the Seafood Community on Errors and Distortions in New Coverage," Aboutseafood.com, accessed August 3, 2012, www.aboutseafood.com/press/open-letter -journalists.

28   J. Hocevar, "NFI Off Base on Pollock" (letter), SeafoodSource.com, published September 17, 2009, http://www.seafoodsource.com/newsarticledetail .aspx?id=9999.

29   "Overfishing of Pollock Risks Collapse of World's Largest Food Fishery, Endangers Sea Lions and Seal," Greenpeace, published December 2, 2008, http://www.greenpeace.org/usa/en/media-center/news-releases/ overfishing-of-pollock-risks-c/.

30   "Too Little, Too Late for Largest US Fishery," Greenpeace, published October 1, 2009, http://www.greenpeace.org/usa/en/news-and-blogs/news/too -little-too-late-for-large/.

31   "The Bering Sea/Aleutian Islands Pollock Fishery," At-Sea Processors As-

sociation, published 2004, http://www.atsea.org/images/pollock_fishery
_description04.pdf.

32    John Sabella, "Taking the First Step: On the Road to Americanization," *Pacific Fishing*, October 1985.

33    Todd Campbell, "Working Conditions on Factory Ships under Scrutiny," *National Fisherman*, July 1991.

CHAPTER TEN

1    Touza, "Fishing on the Far Side of the World."
2    Weber, *From Abundance to Scarcity*.
3    Hornnes, "Norwegian Investments."
4    The *Northern Boobie*?
5    Ramsey, "The Tide of Change."
6    Hornnes, "Norwegian Investments."
7    "Factory Boats Sent to Russian Waters," *National Fisherman*, October 1997, 11–12.
8    It was sold to Trident Seafoods in 1991 as the *Independence*.
9    "Bender Shipyards Build 5 Vessels," *Fishing News International*, April 1985, 4.
10    "Arctic Alaska Looks to New Waters," *Fishing News International*, April 1992, 32.
11    Hal Bernton, "Chuck Bundrant Rides the High Seas of the Seafood Industry," *Seattle Times*, June 24, 2002; also Plesha.
12    "Surimi by Unisea Planned for Dutch Harbor," *Pacific Fishing*, August 1985, 16.
13    "Seattle Will Make Its Crab Meat," *Fishing News International*, May 1985, 13.
14    It was originally the Pacific Pearl Seafoods Plant, and later became part of the larger UniSea operation.
15    David Schaefer and Duff Wilson, "Fish Quota Favors Alaska over Seattle Ships—Critics Call Ruling 'Blatant Politics,'" *Seattle Times*, March 5, 1992.
16    L. Helm, "Catch as Catch Can . . . Seattle's Factory Trawlers Run into an Alaskan Storm," *Seattle Post Intelligencer*, Oct. 23, 1989.
17    North Pacific Fisheries Management Council, "Celebrating 30 Years of Sustainable Fisheries" (Anchorage: North Pacific Fisheries Management Council, 2006).
18    Pereyra.
19    "Pollock Fleet Boom Brings Big Business," *Fishing News International*, December 1988, 7–10.
20    Joel Gay, "Inshore Offshore," *Pacific Fishing*, March 1990, 60–63.
21    Ibid.
22    Brad Warren, "New Twist on Groundfish Wars," *National Fisherman*, August 1992, 11–13.
23    Ross Anderson, "American Factory Trawlers—Fishermen Scramble for Their Piece of the Pollock Pie," *Seattle Times*, June 17, 1991.
24    "We're Out at the Stroke of a Pen," *Fishing News International*, October 1991, 9.

25   Anderson, "American Factory Trawlers."

26   "Factories Hit Back," *Fishing News International*, September 1991, 1.

27   Helm, "Catch as Catch Can."

28   The major US-owned companies with shoreside facilities, Trident and Icicle Seafoods, have their headquarters in Seattle.

29   Brad Matson, "New Politics in the 'Owner State,'" *National Fisherman*, March 1991, 12–16.

30   Ibid.

31   Anderson, "American Factory Trawlers."

32   Ross Anderson, "Fishy Business—Seattle Fishermen Urge Using Market to Allocate Resource," *Seattle Times*, November 10, 1991.

33   Ibid.

34   Matson, "New Politics in the 'Owner State.'"

35   The pollock CDQ was later joined by separate CDQs for halibut and sablefish (1995), other groundfishes (1998), and crab (1998).

36   "Owners Starting Bering Sea Fund," *Fishing News International*, August 1991, 64.

37   Alex DeMarban, "Coastal Villages Catch Fishery Wave," *First Alaskans* (Oct-Nov 2008), http://www.firstalaskansmagazine.com/index.php?issue=10 -2008&story=fishery_wave.

38   Matson, "New Politics in the 'Owner State.'"

39   Schaefer and Wilson, "Fish Quota Favors Alaska over Seattle Ships."

40   Ibid.

41   Tom Brown, "Onshore Plants File Legal Brief in Pollock Battle," *Seattle Times*, June 11, 1992.

42   Haberstroh, "Tyson Goes Fishin'."

43   "Marriage Made in Heaven," *Fishing News International*, October 1992, 20–22.

44   Robert D. McFadden, "Donald J. Tyson, Food Tycoon, Is Dead at 80," *New York Times*, January 8, 2007, 1.

45   Hal Bernton, "Battle for the Deep," *Mother Jones Magazine*, August 1994, 32–38.

46   Ibid.

47   Haberstroh, "Tyson Goes Fishin'."

48   Weber, *From Abundance to Scarcity*.

49   Scott Sunde, "Factory Trawler Fleet May Sit in Port Several Months," *Seattle Post-Intelligencer*, February 26, 1993.

50   "Trawlers: Allocation of Pollock Ruining Us," *Seattle Times*, November 11, 1994.

51   Anderson, "Big Factory Trawlers: To Be or Not To Be?"

52   Jan Jacobs, April 11, 2011.

53   Hal Bernton, "Troubled Waters: Norway Bank Asking Owners of Vessels to Pledge Fishing Rights off Alaska as Collateral," *Anchorage Daily News*, May 1, 1994.

54   "Trawlers: Allocation of Pollock Ruining Us."

55   Weber, *From Abundance to Scarcity*.

56   David Whitney, "Battle for the Bering Sea—Stevens Says Bill Would Protect

Fishery, American Interests," *Anchorage Daily News*, December 14, 1997; also "Senator Seeks Limits on Fleet."

57  Mark Buckley and Brad Warren, "Factory Trawler Bill Splits Fleet," *Pacfic Fishing*, May 1998, 14–15.

58  Subcommittee on Fisheries Conservation, Wildlife, and Oceans, *Oversight Hearing on United States Ownership of Fishing Vessels* (Washington, DC: Government Printing Office, 1998).

59  Bruce Ramsey, "Shore Fishing Plants vs. Trawlers—The Battle Gets Nasty," *Seattle Times*, October 1, 1997.

60  Brad Warren, "How the Council Deals Out Pollock Options," *Pacific Fishing*, November 1997, 33.

61  Bernton, "Bundrant Rides the High Seas."

62  Laura Onstot, "King of Fish Sticks," *Seattle Weekly*, November 19, 2008.

63  Bernton, "Bundrant Rides the High Seas."

64  Mary Lenz, "Dutch Harbor's Tales of Big Bucks," *Lewiston Journal*, January 24, 1980.

65  Weber, *From Abundance to Scarcity*.

66  Onstot, "King of Fish Sticks."

67  Bernton, "Bundrant Rides the High Seas."

68  Ibid.

69  Plesha; Jacobs.

70  B. Mansfield, "Neoliberalism in the Oceans: 'Rationalization,' Property Rights, and the Commons Question," *Geoforum* 35 (2004): 313-26.

71  Ibid.

72  Buckley and Warren, "Factory Trawler Bill Splits Fleet."

73  Ben Speiss, "Peace in the Pollock Business?" *Pacific Fishing*, November 1998, 30–32.

74  "Senator Seeks Limits on Fleet."

75  Buckley and Warren, "Factory Trawler Bill Splits Fleet."

76  Ibid.

77  Ben Speiss, "*Triumph* to Trump Critics," *Pacific Fishing*, August 1998, 32.

78  Onstot, "King of Fish Sticks."

79  Bernton, "Battle for the Deep."

80  Wesley Loy, "The Race for Pollock Ends: Co-Op Ends Competitive Bering Sea Fishery, Improves Safety and Efficiency, but Not Everybody Is Pleased with the New Era," *Anchorage Daily News*, July 30, 2000.

81  Ibid.

82  Ibid.

83  Hal Bernton, "Congress Cedes Fishing Rights to Group," *Seattle Times*, November 19, 2001.

84  Kim Murphy, "Bering Sea: Where Few Even Dare," *Kitsap Sun*, April 28, 2001.

85  Mike Lewis, "Alaska's Wild, Woolly Bar Scene Has Calmed in Recent Years," *Seattle Post-Intelligencer*, October 29, 2003.

86  Ibid.

87 "The Elbow Room," http://www.arctic.net/~elbowrm/.

88 Joel Gay, "High Noon in Alaska," *Pacific Fishing*, February 1993.

89 KUCB, 2007.

90 Gay, "High Noon in Alaska."

91 Bruce Ramsey, "Shore Fishing Plants vs. Trawlers—The Battle Gets Nasty," *Seattle Post-Intelligencer*, October 1, 1997.


CHAPTER ELEVEN

1 Charlie Ess, "Learn and Live: Linda Behnken, Sitka, Alaska," *National Fisherman*, December 2009, 32–34.

2 Linda Behnken, March 17, 2011.

3 PEW Oceans Commission, "A Dialogue on America's Fisheries."

4 Loy, "The Race for Pollock Ends."

5 Rod Fujita, February 1, 2011.

6 A. M. Eikeset et al., "Unintended Consquences Sneak in the Back Door: Making Wise Use of Regulations in Fisheries Management," in *Ecosystem-Based Management for Marine Fisheries: An Evolving Perspective*, ed. A. Belgrano and C. W. Fowler (New York: Cambridge University Press, 2011), 183–217.

7 Mansfield, "Neoliberalism in the Oceans."

8 D. D. Huppert, "An Overview of Fishing Rights," *Reviews in Fish Biology and Fisheries* 15 (2005): 201–15.

9 Mansfield, "Neoliberalism in the Oceans."

10 Weber, *From Abundance to Scarcity*.

11 Ramsey, "The Tide of Change."

12 Ibid. Dan Huppert quoted therein.

13 "Tillion Leaves Council," *Pacific Fishing*, August 1997, 15.

14 Huppert, "An Overview of Fishing Rights."

15 R. Sharp and U. R. Sumaila, "Quantification of US Marine Fisheries Subsidies," *North American Journal of Fisheries Management* 29, no. 1 (Feb 2009): 18–32.

16 "Milestones in the Western Alaska CDQ Program," Coastal Villages, accessed August 3, 2012, http://www.coastalvillages.org/about-us/history.

17 DeMarban, "Coastal Villages Catch Fishery Wave."

18 "Alaska Committee Recommends CDQ Reforms," *Pacific Fishing*, November 2005, 12.

19 Wesley Loy, "CDQ Groups Gain Clout, Controversy," *Pacific Fishing*, January 2003, 10–11.

20 Wesley Loy, "Adak Reaches for Pollock," *Pacific Fishing*, November 2003, 11–12.

21 Wesley Loy, "Congress Steps In to Settle CDQ War," *Pacific Fishing*, June 2006, 6–11.

22 Andrew Jensen, "CDQ Groups Pursue Tax Break even as Some Say They Don't Need It," *Alaska Journal of Commerce*, August 6, 2010.

23    US Census Bureau, "State and County Quick Facts," 2010, http://quickfacts
      .census.gov/qfd/index.html.
24    Western Alaska Community Development Association, *Western Alaska Com-
      munity Development Quota Program* (Anchorage: Western Alaska Community
      Development Association, 2010).
25    Andrew Jensen, "2 CDQ Execs Reel In Nearly $4 Million since 2006," *Alaska
      Journal of Commerce*, August 6, 2010.
26    Andrew Jensen, "Washington, Oregon Crab Interests Take Aim at CDQs,
      Alaska Council Majority," *Alaska Journal of Commerce*, September 15, 2011.
27    S. Hanna et al., *Fishing Grounds* (Washington, DC: Island Press, 2000).
28    Hal Bernton, "Alaska, Washington in Fishing-Fleet Tug of War," *Seattle
      Times*, October 1, 2011.
29    Ibid.
30    Ross Anderson, "Pollock Politics."
31    Ross Anderson and Duff Wilson, "A Fishy Situation—Critics Say Members
      of Panel Set Up to Manage Rich Fisheries Zone off Alaska Are Watching over
      Their Own Self-Interests at the Same Time," *Seattle Times*, November 10, 1991.
32    Anderson, "Fishy Business."
33    Anderson and Wilson, "A Fishy Situation."
34    Anderson, "Fishy Business."
35    Anderson and Wilson, "A Fishy Situation."
36    Anderson, "Fishy Business."
37    Hanna et al., *Fishing Grounds*.
38    Anderson and Wilson, "A Fishy Situation"; also "Conflict of Interest Stan-
      dards of Regional Fishery Management Councils," Institute of the North,
      published 2008, http://www.institutenorth.org/assets/images/uploads/files/
      MCA_CI.
39    "Magnuson-Stevens Fishery Conservation and Management Act; Regional
      Fishery Management Councils; Operations" (Washington, DC: Federal Re-
      gister, 2010), https://www.federalregister.gov/articles/2010/09/27/2010
      -24222/magnuson-stevens-fishery-conservation-and-management-act
      -regional-fishery-management-councils#p-3.
40    "Conflict of Interest Standards."
41    R. Hilborn, "Pretty Good Yield and Exploited Fishes," *Marine Policy* 34 (2010):
      193–96.
42    C. W. Fowler and S. M. McClusky, "Sustainability, Ecosystems, and Fish-
      ery Management," in *Ecosystem-Based Management for Marine Fisheries: An
      Evolving Perspective*, ed. A. Belgrano and C.W. Fowler (New York: Cambridge
      University Press, 2011), 307–36.
43    Greenberg, *Four Fish*.
44    Eikeset et al., "Unintended Consquences Sneak in the Back Door."
45    M. Lima, "Populations Dynamics Theory as an Essential Tool for Models in
      Fisheries," in *Ecosystem-Based Management for Marine Fisheries: An Evolving*

*Perspective*, ed. A. Belgrano and C.W. Fowler (New York: Cambridge University Press, 2011), 218–31.

46  Anderson, "Big Factory Trawlers: To Be or Not to Be?"

47  Clover, *End of the Line*. Holt cited therein.

48  Hanna et al., *Fishing Grounds*.

49  Clover, *End of the Line*.

50  Ramsey, "The Tide of Change." Bill Aron, quoted therein.

51  Mueter et al., "Expected Declines in Recruitment of Walleye Pollock."

52  C. Hsieh et al., "Fishing Effects on Age and Spatial Structures Undermine Population Stability of Fishes," *Aquatic Sciences* 72 (2010): 165–78.

53  Lewis, "Alaska's Wild, Woolly Bar Scene Has Calmed."

54  Wesley Loy, "The Brig: Dutch Harbor Report," *The Deckboss* (2011), http://deckboss-thebrig.blogspot.com/2011/07/dutch-harbor-report_18.html.

CHAPTER TWELVE

1   Hal Bernton, "Pollock Stocks Expected to Be Strong," *Seattle Times*, November 25, 2010.

2   Eric Hoffner, "Most Ubiquitous Fish in American Diet 50 Percent below Last Year's Levels," in Grist, published October 14, 2008, http://www.grist.org/article/pollock-poster-fishery-on-the-brink/PALL/print.

3   Frank Kelty, "Unalaska Fisheries Update, August 29, 2011," published August 29, 2011, http://unalaska-ak.us/vertical/Sites/%7B0227B6A7-A82F-4BFC-9D02-A4B2D3A8BC35%7D/uploads/Fisheries_Update_and_NMFS_Catch_Reports_Aug_29_2011.pdf.

4   This issue is the topic of the PhD dissertation of Stan Kotwicki, a University of Washington graduate student.

5   Andrew Jensen, "Health of Pollock Stock Debated, Quota Set," *Alaska Journal of Commerce*, December 15, 2011.

6   Ibid.

7   Andrew Jensen, "Trident Gears Up for Snow Crab, Responds to Pollock Chatter," *Alaska Journal of Commerce*, January 4, 2012.

8   Helen Jung, "Cashing In on Pollock—$650 Million Pollock Fishery Thrives in Bering Sea," *Anchorage Daily News*, April 12, 1998.

9   Ibid.

10  Pedersen was a crab fisherman with interests in five or six crab boats. In 1978 he bought a retired AOG-1 Navy gasoline tanker for $400,000 with his son Mark, Stan Hovik, and Martin Stone, intending to convert it to a processor, but they were unable to get financing (Arestad, "Norwegians in the Pacific Coast Fisheries"). Wally Pereyra's company ProFish, which had invested in the ship with the original owners, bought it and worked with the Norwegian marine architect Lars Eldøy to get financing from Christiana Bank of Norway and equity from Korean investors for the project. They converted the ship into the 340-foot surimi factory trawler *Arctic Storm*.

11   J. N. Ianelli et al., "Assessment of the Walleye Pollock Stock in the Eastern Bering Sea," *Stock Assessment and Fisheries Evaluation Report* (2009): 1–136, http://www.afsc.noaa.gov/refm/docs/2009/EBSpollock.pdf.

12   R. Beverton, "Fish, Fact, and Fantasy: A Long View," *Fish and Fisheries* 8 (1998): 229–49.

13   Charlie Ess, "Strict Controls Earn Ecolabel Renewal; Higher Quota Seen as Possible for 2011," *National Fisherman*, November 2010.

14   Frank Kelty, "Unalaska Fisheries Update, July 16, 2012," published July 16, 2012, http://www.unalaska-ak.us/vertical/sites/%7B0227B6A7-A82F-4BFC-9D02-A4B2D3A8BC35%7D/uploads/Fisheries_update_July_16_2012.pdf.

# Bibliography

"9 Arrested in Protest of Factory Fishing." *Seattle Times*, August 2, 1995.

"Acona: Surimi and Fillets from Alaska to New Zealand." *Fishing News International*, July 1990, 44–49.

Akira, N., T. Yanagimoto, K. Mito, and S. Katakura. "Interannual Variability in Growth of Walleye Pollock, *Theragra Chalcogramma*, in the Central Bering Sea." *Fisheries Oceanography* 10 (2001): 367–75.

"Alaska Committee Recommends CDQ Reforms." *Pacific Fishing*, November 2005, 12.

"Alaska Pollock–Japanese Pacific." FishSource. Published 2010. http://www .fishsource.org/fishery/identification?fishery=Alaska+pollock+-+Japanese+ Pacific.

"Alaska Pollock–Sea of Okhotsk." FishSource. Published 2010. http://www .fishsource.com/fishery/identification?fishery=Alaska+pollock+-+Sea+ of+Okhotsk+%28Country%3A+RU%3B+Gear%3A+TM%3B+MSC-Client% 3A+RPA%3B+MSC-Status%3A+MSC+Full+Assessment%3B%29.

"Alaska Pollock–West Bering Sea." FishSource. Published 2010. http://www .fishsource.org/fishery/data_summary?fishery=Alaska+pollock+-+W+ Bering+Sea.

Alverson, D. L. "Technical Report 1: The Pacific Northwest Bottomfish Resources and Fisheries—a Historical Perspective." In *A Strategy for the Americanization of the Groundfish Fisheries of the Northeast Pacific*. Saltonstall-Kennedy Grant Program. Seattle: Pacific Seafood Processors Association, 1985.

Alverson, D. L. February 18, 2011.

Alverson, D. L. *Race to the Sea*. New York: iUniverse, 2008.

Alverson, D. L. September 12, 2007.

Ames, E. P. "Atlantic Cod Stock Structure in the Gulf of Maine." *Fisheries* 29, no. 1 (2004): 10–28.

Anda, Torgeir. *Røkke*. Oslo: Universitetsforlaget, 1997.

Anderson, Ross. "American Factory Trawlers—Fishermen Scramble for Their Piece of the Pollock Pie." *Seattle Times*, June 17, 1991.

Anderson, Ross. "Big Factory Trawlers: To Be or Not to Be? Hearings Begin Tomorrow on Possible Ban of Vessels." *Seattle Times*, March 25, 1998.

Anderson, Ross. "Factory Trawlers Facing Rough Seas–Greenpeace, Political Foes Line Up against Big Boats." *Seattle Times*, January 9, 1998.

Anderson, Ross. "Fishy Business—Seattle Fishermen Urge Using Market to Allocate Resource." *Seattle Times*, November 10, 1991.

Anderson, Ross. "The Politics of Pollock." *Seattle Times*, February 12, 1991.

Anderson, Ross. "Pollock Politics: A Billion-Dollar Tragedy." *Seattle Times*, August 14, 1992.

Anderson, Ross, and Duff Wilson. "A Fishy Situation—Critics Say Members of Panel Set Up to Manage Rich Fisheries Zone off Alaska Are Watching over Their Own Self-Interests at the Same Time." *Seattle Times*, November 10, 1991.

Andrewartha, H. G., and L. C. Birch. *The Distribution and Abundance of Animals.* Chicago: University of Chicago Press, 1954.

"Arctic Alaska Looks to New Waters." *Fishing News International*, April 1992, 32.

Arestad, Sverre. "Norwegians in the Pacific Coast Fisheries." *Norwegian American Studies* 30 (1985): 13.

Armitage, D., ed. *Hugo Grotius, The Free Sea, Trans. Richard Hakluyt, with William Welwod's Critique and Grotius's Reply [1609].* Indianapolis: Liberty Fund, 2004. http://oll.libertyfund.org/title/859/66149.

Associated Press. "Fishing Restrictions to Aid Sea Lions Target of Hearing." *Seattle Times*, October 17, 2011.

Bailey, K. M., L. Ciannelli, N. A. Bond, A. Belgrano, and N. C. Stenseth. "Recruitment of Walleye Pollock in a Physically and Biologically Complex Ecosystem: A New Perspective." *Progress in Oceanography* 67, no. 1–2 (2005): 24–42.

Bailey, K. M., T. J. Quinn, P. Bentzen, and W. S. Grant. "Population Structure and Dynamics of Walleye Pollock, *Theragra Chalcogramma.*" *Advances in Marine Biology* 37 (2000): 179–255.

Baird, R. "Illegal, Unreported, and Unregulated Fishing: An Analysis of the Legal, Economic, and Historical Factors Relevant to Its Development and Persistence." *Melbourne Journal of International Law* 5 (2004): 299–334. http://www.austlii.edu.au/au/journals/MelbJIL/2004/13.html.

Balykin, P. A., and A. V. Buslov. "Long-Term Variability in Length of Walleye Pollock in the Western Bering Sea and East Kamchatka." *PICES Scientific Report* 20 (2002): 67–69.

Behnken, Linda. March 17, 2011.

"Bender Shipyards Build 5 Vessels." *Fishing News International*, April 1985, 4.

Berglund, Nina. "Labour's Capitalist Darling Pushes His Luck." News in English. Published April 21, 2009. http://www.newsinenglish.no/2009/04/21/labour%E2%80%99s-capitalist-darling-pushes-his-luck/.

"The Bering Sea/Aleutian Islands Pollock Fishery." At-Sea Processors Association. Published 3, 2004. http://www.atsea.org/images/pollock_fishery_description04.pdf.

Bernton, Hal. "Alaska, Washington in Fishing-Fleet Tug of War." *Seattle Times*, October 1, 2011.

Bernton, Hal. "Battle for the Deep." *Mother Jones Magazine*, August 1994, 32–38.

Bernton, Hal. "Chuck Bundrant Rides the High Seas of the Seafood Industry." *Seattle Times*, June 24, 2002.

Bernton, Hal. "Congress Cedes Fishing Rights to Group." *Seattle Times*, November 19, 2001.

Bernton, Hal. "Pollock Stocks Expected to Be Strong." *Seattle Times*, November 25, 2010.

Bernton, Hal. "Seattle Trawlers May Face New Limits on Crucial Pollock Fishery." *Seattle Times*, October 11, 2008.

Bernton, Hal. "Troubled Waters: Norway Bank Asking Owners of Vessels to Pledge Fishing Rights off Alaska as Collateral." *Anchorage Daily News*, May 1, 1994.

Bernton, Hal. "Will Seafood Nets Be Empty? Grim Outlook Draws Skeptics." *Seattle Times*. November 3, 2006.

Beverton, R. "Fish, Fact, and Fantasy: A Long View." *Fish and Fisheries* 8 (1998): 229–49.

Beverton, R. J. H., and S. J. Holt. *On the Dynamics of Exploited Fish Populations: Fisheries Investigations Series II*. London: Ministry of Agriculture, Fisheries, and Food, 1957.

Bird, Chris. "Russia's Favorite Fish on Verge of Extinction." *New Scientist* 140, no. 1902 (1993): 10.

Bluemink, E. "Herring Fishermen Strike Mother Lode off Alaska." *Anchorage Daily News*, March 24, 2008.

Bodal, Bernt. February 9, 2011.

Bottom, D. L. "To Till the Water: A History of Ideas in Fisheries Conservation." In *Pacific Salmon & Their Ecosystems: Status and Future Options*, edited by D. J. Stouder, P. A. Bisson, and R. J. Naiman, 569–97. Detroit: Chapman and Hill, 1997.

Branch, T. A., O. P. Jensen, D. Ricard, Y. M. Ye, and R. Hilborn. "Contrasting Global Trends in Marine Fishery Status Obtained from Catches and from Stock Assessments." *Conservation Biology* 25, no. 4 (Aug 2011): 777–86.

Branch, T. A., Reg Watson, Elizabeth A. Fulton, et al. "The Trophic Fingerprint of Marine Fisheries." *Nature* 468, no. 7322 (2010): 431–35.

Breivik, Erik. November 14, 2011.

Brodeur, R. D., M. B. Decker, L. Ciannelli, et al. "Rise and Fall of Jellyfish in the Eastern Bering Sea in Relation to Climate Regime Shifts." *Progress in Oceanography* 77, no. 2–3 (May-Jun 2008): 103–11.

Brooks, P. M., and U. R. Sumaila. "Without Drastic Measures, Gulf of Maine Cod Fishery Will Be Lost Forever." *Portland Press Herald/Maine Sunday Telegram*, April 2, 2006. http://www.seaaroundus.org/newspapers/2006/MaineSunday Telegram_April2006.pdf.

Brown, Tom. "Onshore Plants File Legal Brief in Pollock Battle." *Seattle Times*, June 11, 1992.

Buckley, Mark, and Brad Warren. "Factory Trawler Bill Splits Fleet." *Pacfic Fishing*, May 1998, 14–15.

Bull, Francis. *Norske Portretter: Videnskapsmenn (Norwegian Portraits: Scientists)*. Copenhagen: Gyldendal, 1965.

Burkenroad, M. D. "Some Principles of Marine Fishery Biology." *Publ Inst Mar Sci Univ Tex* 2 (1951): 177-212.

Butcher, David. *The Trawlermen*. Reading, United Kingdom: Tops'l Books, 1980.

Campbell, Todd. "Working Conditions on Factory Ships under Scrutiny." *National Fisherman*, July 1991, 11-14.

Carson, Rachel. "Food from the Sea: Fish and Shellfish from New England." *Conservation Bulletin*, 1-74. Washington, DC: US Department of the Interior, US Fish and Wildlife Service, 1943.

Carson, Rachel. *Silent Spring*. Boston: Houghton-Mifflin, 1962.

Carson, Rachel. *The Sea around Us*. New York: Oxford University Press, 1950.

Cavanagh, Sean. "Greenpeace Ends Bridge Protest—Weekend Effort Turns Back Fishing Trawler." *Seattle Times*, August 18, 1997.

CBS 2007: 12th Annual Conference of the Parties to the Convention on the Conservation and Management of Pollock Resources in the Central Bering Sea. Virtual conference hosted by China. http://www.afsc.noaa.gov/REFM/CBS/Docs/12th%20Annual%20Conference/12th%20Annual%20Donut%20Hole%20Final%20Report.pdf.

CBS 2010: 15th Annual Conference of the Parties to the Convention on the Conservation and Management of Pollock Resources in the Central Bering Sea. Virtual conference hosted by the United States. http://www.afsc.noaa.gov/refm/cbs/Docs/15th%20Annual%20Conference/Final%20Report%20of%20the%2015th%20AC%2010-6_edt.pdf.

"Centuries of Fish: Seattle's Dynamic High Seas Fisheries Fleet" (film), produced by Bob Thorstenson. Seattle: Maritime Heritage Foundation, 2003.

Christy, F. T., and A. Scott. *The Common Wealth in Ocean Fisheries*. Baltimore: Johns Hopkins University Press, 1965.

"Chronicle of a Plunder Foretold." Published November 17, 1997. http://archive.greenpeace.org/comms/fish/plunsum.html.

Clover, C. *The End of the Line*. London: Ebury Press, 2004.

Collins, J. W. "Report on the Fisheries of the Pacific Coast of the United States." *Report of Commissioner for 1888*, 3-269. Washington, DC: US Commission of Fish and Fisheries, 1892.

"Conflict of Interest Standards of Regional Fishery Management Councils." Institute of the North. Published 2008. http://www.institutenorth.org/partners/studies-and-reports/.

Conover, S., and V. Monakhov. "Russia Review." *Pacific Fishing*, March 1993, 19.

Corten, A. "A Proposed Mechanism for the Bohuslän Herring Periods." *ICES Journal of Marine Science* 56 (1999): 207-20.

Cushing, D. H. *Marine Ecology and Fisheries*. Cambridge: Cambridge University Press, 1975.

Daniel, H., and F. Minot. *The Inexhaustible Sea*. New York: Collier Books, 1961.

Davis, G. "US and Russia Team Up in Bering Sea." *Pacific Fishing*, Yearbook 1994, 28.

DeMarban, Alex. "Coastal Villages Catch Fishery Wave." First Alaskans (Oct-Nov 2008). http://www.firstalaskansmagazine.com/index.php?issue=10 -2008&story=fishery_wave.

DeMaster, D. P., A. W. Trites, P. Clapham, et al. "The Sequential Megafaunal Collapse Hypothesis: Testing with Existing Data." *Progress in Oceanography* 68, no. 2-4 (2006): 329-42.

Dillon, Patrick. *Lost at Sea*. New York: Touchstone, 1998.

"The Doughnut-Hole Pact—Overfished Bering Sea Will Get a Welcome Break." *Seattle Times*, August 27, 1992. http://community.seattletimes.nwsource.com/ archive/?date=19920827&slug=1509661.

Eikeset, A. M., A. P. Richter, F. K. Diekert, D. J. Dankel, and N. C. Stenseth. "Unintended Consequences Sneak in the Back Door: Making Wise Use of Regulations in Fisheries Management." In *Ecosystem-Based Management for Marine Fisheries: An Evolving Perspective*, edited by A. Belgrano and C. W. Fowler, 183-217. New York: Cambridge University Press, 2011.

"The Elbow Room." Accessed August 3, 2012. http://www.arctic.net/~elbowrm/.

Ess, Charlie. "Learn and Live: Linda Behnken, Sitka, Alaska." *National Fisherman*, December 2009, 32-34.

Ess, Charlie. "Strict Controls Earn Ecolabel Renewal; Higher Quota Seen as Possible for 2011." *National Fisherman*, November 2010.

Estes, J. A., M. T. Tinker, T. M. Williams, and D. F. Doak. "Killer Whale Predation on Sea Otters Linking Oceanic and Nearshore Ecosystems." *Science* 282 (1998): 473-76.

"Factories Hit Back." *Fishing News International*, September 1991, 1.

"Factory Boats Sent to Russian Waters." *National Fisherman*, October 1997, 11-12.

Fauchald, P., M. Mauritzen, and H. Gjosaeter. "Density-Dependent Migratory Waves in the Marine Pelagic Ecosystem." *Ecology* 87 (2006): 2915-24.

Finley, M. C. *All the Fish in the Sea: Fish, Fisheries Science, and Foreign Policy*. Chicago: University of Chicago Press, 2011.

Fowler, C.W., and S. M. McClusky. "Sustainability, Ecosystems, and Fishery Management." In *Ecosystem-Based Management for Marine Fisheries: An Evolving Perspective*, edited by A. Belgrano and C.W. Fowler, 307-36. New York: Cambridge University Press, 2011.

Francis, R. C. "Fisheries Science Now and in the Future: A Personal View." *New Zealand Journal of Marine and Freshwater Research* 14 (1980): 95-100.

Frank, K.T., B. Petri, and N. L. Shackell. "The Ups and Downs of Trophic Control in Continental Shelf Ecosystems." *Trends in Ecology and Evolution* 22 (2007): 236-42.

Fraser, Dave. July 9, 2011.

Fredin, R. A. "History of Regulation of Alaska Groundfish Fisheries." *Processed Report*, 63. Seattle: Northwest and Alaska Fisheries Center, 1987.

Friedland, J. "Rivalries Grow for Global Fishers, as Fleets Expand and Hauls Wane." *Wall Street Journal*, interactive edition. November 25, 1997. http://www.econ.ucsb.edu/~tedb/eep/news/fish2.ext.txt

Fritz, L. "Trawl Locations of Walleye Pollock and Atka Mackerel Fisheries in the Bering Sea, Aleutian Islands and Gulf of Alaska from 1977–92." *AFSC Processed Report* 93–08. Seattle: Seattle: Alaska Fisheries Science Center, 1993.

Fritz, L. W., R. C. Ferrero, and R. J. Berg. "The Threatened Status of Steller Sea Lions, *Eumetopias Jubatus*, under the Endangered Species Act: Effects on Alaska Groundfish Fisheries Management." *Marine Fisheries Review* 57, no. 2 (1995): 14–27.

"From Bare Hands to Global Embrace." World Fishing and Aquaculture. Published February 27, 2006. http://www.worldfishing.net/comment-and-analysis101/interviews/from-bare-hands-to-global-embrace.

Fuchs, S. J. "Feeding the Japanese: Food Policy, Land Reform, and Japan's Economic Recovery." In *Democracy in Occupied Japan*, edited by Mark Caprio and Yoneyuki Sugita, 26–47. New York: Routledge, 2007.

Fujita, Rod. February 1, 2011.

Gay, Joel. "Filling the Doughnut Hole." *Pacific Fishing*, November 1992, 24–25.

Gay, Joel. "High Noon in Alaska." *Pacific Fishing*, February 1993, 26–27.

Gay, Joel. "Inshore Offshore." *Pacific Fishing*, March 1990, 60–63.

Gay, Joel. "Will Factory Trawlers Last into the Future?" *Pacific Fishing*, October 1992, 48–56.

Getz, W. M., and R. G. Haight. *Population Harvesting*. Princeton, NJ: Princeton University Press, 1989.

Glantz, M. H. "Science, Politics, and Economics of the Peruvian Anchoveta Fishery." *Marine Policy* 3 (1979): 201–10.

Graham, M. "Modern Theory of Exploiting a Fishery and Applications to North Sea Trawling." *Journal Conseil International pour l'Exploration Mer* 10 (1935): 264–74.

Graham, M. *The Fish Gate*. London: Faber and Faber, 1949.

Greenberg, P. *Four Fish: The Future of the Last Wild Food*. New York: Penguin Press, 2010.

Gregory, H. E., and K. Barnes. *North Pacific Fisheries: With Special Reference to Alaska Salmon*. San Francisco: American Council Institute of Pacific Relations, 1939.

Grotius, H. *Mare Liberum (The Free Sea)*. Oxford: Oxford University Press, 1609.

Guthrie-Shimizu, S. "Occupation Policy and the Japanese Fisheries Management Regime, 1945–1952." In *Democracy in Occupied Japan*, edited by M. E. Caprio and Y. Sugita, 48–66. New York: Routledge, 2007.

Haberstroh, Joe. "Tyson Goes Fishin'— After Arctic Alaska Acquisition, Company Finds Seafood, Chicken Businesses Are Oceans Apart." *Seattle Times*, May 15, 1994.

Hanna, S., H. Blough, R. Allen, S. Iudicello, G. Matlock, and B. McCay. *Fishing Grounds*. Washington, DC: Island Press, 2000.

Hansan, Mary Anne. "Open Letter to Journalists from the Seafood Community on Errors and Distortions in New Coverage." Aboutseafood.com. Accessed August 3, 2012. www.aboutseafood.com/press/open-letter-journalists.

Helm, L. "Catch as Catch Can . . . Seattle's Factory Trawlers Run into an Alaskan Storm." *Seattle Post Intelligencer*, Oct. 23, 1989.

Helvarg, D. "Full Nets, Empty Seas—Harmful Effects of Supertrawlers on Ocean Fisheries." *Progressive*, November 1997.

Hilborn, R. "*Mismanagement of Marine Fisheries* by Alan Longhurst" (review). *Fisheries* 36, no. 5 (2011): 247.

Hilborn, R. "Pretty Good Yield and Exploited Fishes." *Marine Policy* 34 (2010): 193–96.

Hilborn, R. "Reinterpreting the State of Fisheries and Their Management." *Ecosystems* 10, no. 8 (2007): 1362–69.

Hinckley, S. "Spawning Dynamics and Fecundity of Walleye Pollock (*Theragra chalcogramma*) in the Eastern Bering Sea." Master's thesis, University of Washington, 1986.

Hjort, J. "Fluctuations in the Year Classes of Important Food Fishes." *Journal du Conseil* 1 (1926): 5–38.

Hocevar, J. "NFI Off Base on Pollock" (letter). SeafoodSource.com. Published September 17, 2009. http://www.seafoodsource.com/newsarticledetail.aspx?id=9999.

Hoffner, Eric. "Most Ubiquitous Fish in American Diet 50 Percent below Last Year's Levels." In Grist. Published October 14, 2008. http://www.grist.org/article/pollock-poster-fishery-on-the-brink/PALL/print.

"Hole Used for Illegal Fishing." *Fishing News International*, June 1988, 31.

Holt, S. "Three Lumps of Coal: Doing Fisheries Research in Lowestoft during the 1940s." Talk given April 25, 2008. Accessed August 3, 2012. http://www.cefas.defra.gov.uk/media/306484/sidney-holts-tallk.pdf.

Hornnes, R. "Norwegian Investments in the US Factory Trawler Fleet, 1980–2000." Master's thesis, University of Bergen, 2006.

Hsieh, C., A. Yamauchi, T. Nakazawa, and W. Wang. 2010. Fishing Effects on Age and Spatial Structures Undermine Population Stability of Fishes. *Aquatic Sciences* 72: 165–78.

Hubbard, J. W. "The Gospel of Efficiency and the Origins of MSY: Scientific and Social Influences on Johan Hjort and A. G. Huntsman's Contributions to Fisheries Science." Paper presented at the North American Society for Ocean History, Mystic Seaport, Connecticut, May 13, 2010.

Hunt, G.L., and K. F. Drinkwater. "Background on the Climatology, Physical Oceanography, and Ecosytems of the Sub-Arctic Seas: Appendix to the Essas Science Plan." *GLOBEC Report* 96 (2005).

Huppert, D. D. "An Overview of Fishing Rights." *Reviews in Fish Biology and Fisheries* 15 (2005): 201–15.

Huppert, D. D. "Managing the Groundfish Fisheries of Alaska: History and Prospects." *Reviews in Aquatic Sciences* 4, no. 4 (1991): 339–73.

Ianelli, J. N., S. Barbeaux, T. Honkalehto, S. Kotwicki, K. Aydin, and N. William-
    son. "Assessment of the Walleye Pollock Stock in the Eastern Bering Sea." *Stock
    Assessment and Fisheries Evaluation Report* (2009): 1–136. http://www.afsc.noaa
    .gov/refm/docs/2009/EBSpollock.pdf.

Ianelli, J. N., T. Hokalehto, and N. Williamson. "An Age-Structured Assessment of
    Pollock (*Theragra Chalcogramma*) from the Bogoslof Island Region." *Stock As-
    sessment and Fishery Evaluation Report* (2006): 201–36. http://www.afsc.noaa
    .gov/refm/docs/2006/BOGpollock.pdf.

Izhevskii, G. I. *Forecasting of Oceanological Conditions and the Reproduction of Com-
    mercial Fish*. Translated by Israel Program for Scientific Translations. Moscow:
    All-Union Scientific Research Institute of Marine Fisheries and Oceanogra-
    phy, 1964.

Jackson, R. I., and W. F. Royce. *Ocean Forum: An Interpretive History of the Inter-
    national North Pacific Fisheries Commission*. Farnham, England: Fishing News,
    1986.

Jacobs, Jan. April 11, 2011.

Jelley, Susan. "The Price of Fishing in American Waters." *Fishing News Interna-
    tional*, May 1983, 8–9.

Jensen, Andrew. "2 CDQ Execs Reel In Nearly $4 Million since 2006." *Alaska Jour-
    nal of Commerce*, August 6, 2010.

Jensen, Andrew. "CDQ Groups Pursue Tax Break even as Some Say They Don't
    Need It." *Alaska Journal of Commerce*, August 6, 2010.

Jensen, Andrew. "Health of Pollock Stock Debated, Quota Set." *Alaska Journal of
    Commerce*, December 15, 2011.

Jensen, Andrew. "Trident Gears Up for Snow Crab, Responds to Pollock Chatter."
    *Alaska Journal of Commerce*, January 4, 2012.

Jensen, Andrew. "Washington, Oregon Crab Interests Take Aim at CDQs, Alaska
    Council Majority." *Alaska Journal of Commerce*, September 15, 2011.

Josephson, P. R. *Industrialized Nature*. Washington, DC: Island Press, 2002.

Jung, Helen. "Cashing In on Pollock—$650 Million Pollock Fishery Thrives in Ber-
    ing Sea." *Anchorage Daily News*, April 12, 1998.

Kalland, A. *Fishing Villages in Tokugawa Japan*. Honolulu: University of Hawaii
    Press, 1995.

Kasahara, H. *Fisheries Resource of the North Pacific Ocean, Part 1: A Series of Lectures
    Presented at the University of British Columbia, January and February, 1960*. Van-
    couver: University of British Columbia, 1961.

Kasahara, H., and W. Burke. "North Pacific Fisheries Management." In *The Pro-
    gram of International Studies of Fishery Arrangements*, 1–91. Washington, DC:
    Resources for the Future, 1973.

Kelty, Frank. "Unalaska Fisheries Update, August 29, 2011." Published August 29,
    2011. http://www.unalaska-ak.us/index.asp?Type=B_BASIC&SEC={25EBD
    60B-5C2D-4B53-A9D8-D8DC3E122E03}.

Kelty, Frank. "Unalaska Fisheries Update, July 16, 2012." Published July 16, 2012.

http://www.unalaska-ak.us/index.asp?Type=B_BASIC&SEC={25EBD60B
-5C2D-4B53-A9D8-D8DC3E122E03}.

Kendall, A. W., and G. J. Duker. "The Development of Recruitment Fisheries
Oceanography in the United States." *Fisheries Oceanography* 7 (1998): 69–88.

Knight, D. "Super Trawler Threatens Marine Food Chain." Inter Press Service.
Published January 13, 1998. http://www.ips.fi/koulut/199803/11.htm.

Koslow, Anthony. *The Silent Deep*. Chicago: University of Chicago Press, 2007.

Kotenev, B. N., and O. A. Bulatov. "Dynamics of the Walleye Pollock Biomass in the
Sea of Okhotsk." *PICES Scientific Report* 36 (2009): 291–95.

KUCB. 2007. http://kucb.org.

Larkin, P. A. "An Epitaph for the Concept of Maximum Sustainable Yield." *Transactions of the American Fisheries Society* 106 (1977): 1–11.

Lenz, Mary. "Dutch Harbor's Tales of Big Bucks." *Lewiston Journal*, January 24,
1980.

Lewis, Mike. "Alaska's Wild, Woolly Bar Scene Has Calmed in Recent Years."
*Seattle Post-Intelligencer*, Wednesday, October 29, 2003.

Lima, M. "Populations Dynamics Theory as an Essential Tool for Models in
Fisheries." In *Ecosystem-Based Management for Marine Fisheries: An Evolving Per-
spective*, edited by A. Belgrano and C.W. Fowler, 218–31. New York: Cambridge
University Press, 2011.

Lloyd, Graham. "No-Go Scaremongers 'Fishing for Funds.'" *Australian*, March 1,
2012.

Loew, Chris. "Surimi Import Prices Level Off in Japan." In *SeafoodSource.com*. Published November 2, 2009. http://seafoodsource.com/MarketReport.aspx?id
=4294976522.

Longhurst, A. R. *Mismanagement of Marine Fisheries*. New York: Cambridge University Press, 2010.

Lotka, A. *Elements of Mathematical Biology*. New York: Dover, 1956. Reprint of *Elements of Physical Biology*. Baltimore: Williams and Wilkins, 1925.

Loy, Wesley. "Adak Reaches for Pollock." *Pacific Fishing*, November 2003, 11–12.

Loy, Wesley. "CDQ Groups Gain Clout, Controversy." *Pacific Fishing*, January 2003,
10–11.

Loy, Wesley. "Congress Steps In to Settle CDQ War." *Pacific Fishing*, June 2006, 6–11.

Loy, Wesley. "Making Waves—How Bernt Bodal Navigated His Company to the
Lead of the Pollock Fishery." *Anchorage Daily News*, September 11, 2005.

Loy, Wesley. "The Brig: Dutch Harbor Report." In *The Deckboss*. Published July 18,
2011. http://deckboss-thebrig.blogspot.com/2011/07/dutch-harbor-report_18.
html.

Loy, Wesley. "The Race for Pollock Ends: Co-Op Ends Competitive Bering Sea
Fishery, Improves Safety and Efficiency, but Not Everybody Is Pleased with the
New Era." *Anchorage Daily News*, July 30, 2000.

Loy, Wesley. "War Horse: Clem Tillion Scores Another Victory at Adak." *Pacific
Fishing*, August 2004, 31–32.

MacCall, A. D. *Dynamic Geography of Marine Fish Populations*. Seattle: University of Washington Press, 1990.

Madslien, Jorn. "Norway Tycoon Gets Jail for Boat Crime." In *BBC News*. Published July 1, 2005. http://news.bbc.co.uk/2/hi/business/4641781.stm.

"Magnuson-Stevens Fishery Conservation and Management Act; Regional Fishery Management Councils; Operations." Washington, DC: Federal Register, 2010. https://www.federalregister.gov/articles/2010/09/27/2010-24222/magnuson-stevens-fishery-conservation-and-management-act-regional-fishery-management-councils#p-3.

Malmin, Oyvind. "Norwegian Americans in the King Crab Fishery." Master's thesis, University of Bergen. 2008.

Mangelsdorf, T. *History of Steinbeck's Cannery Row*. Santa Cruz: Western Tanager Press, 1986.

Mansfield, B. "Neoliberalism in the Oceans: 'Rationalization,' Property Rights, and the Commons Question." *Geoforum* 35 (2004): 313-26.

Marine Stewardship Council. http://www.msc.org/.

"Marriage Made in Heaven." *Fishing News International*, October 1992, 20-22.

Mathiesen, O., and D. E. Bevan. *Some International Aspects of Soviet Fisheries*. Columbus: Ohio State University Press, 1968.

Matson, Brad. "New Politics in the 'Owner State.'" *National Fisherman*, March 1991, 12-16.

Matsuda, Y. "History of Fisheries Science in Japan." In *Oceanographic History: The Pacific and Beyond*, edited by K. R. Benson and P. F. Rehbock, 405-16. Seattle: University of Washington Press, 1993.

McBeath, J. "Greenpeace v. National Marine Fisheries Service: Steller Sea Lions and the Commercial Fisheries in the North Pacific." *Alaska Law Review* 21 (2004): 1-42.

McClure, Robert. "Jellyfish for Lunch? It's No Joke, Says Scientist." *Seattle Post-Intelligencer*, May 3, 2004.

McEvoy, A. F. *The Fisherman's Problem*. Cambridge: Cambridge University Press, 1986.

McFadden, Robert D. "Donald J. Tyson, Food Tycoon, Is Dead at 80." *New York Times*, January 8, 2007, 1.

McGinley, Beth. "Erik Breivik." *Pacific Fishing*, December 1988, 50-55.

"Megatrawl Developed for Pollock Fishery." *Fishing News International*, March 1989, 82-83.

Megrey, Bernard A., and V. G. Wespestad. "Alaska Groundfish Resources: 10 Years of Management under MGCMA." *North American Journal of Fisheries Management* 10, no. 2 (1990): 125-43.

Merrick, R. L., T. R. Loughlin, and D. G. Calkins. "Decline in Abundance of the Northern Sea Lion, *Eumetopias Jubatus*, in Alaska." *Fishery Bulletin* 85 (1987): 351-65.

Miles, E. April 14, 2011.

Miles, E. L. *The US/Japan Fisheries Relationship in the Northeast Pacific: From Conflict to Cooperation?* Seattle: Fisheries Management Foundation and Fisheries Research Institute, 1989.

Miles, E., S. Gibbs, D. Fluharty, C. Dawson, and D. Teeter. *The Management Regime of Marine Regions: The North Pacific.* Berkeley: University of California Press, 1982.

"Milestones in the Western Alaska CDQ Program." Coastal Villages. Accessed August 3, 2012. http://www.coastalvillages.org/about-us/history.

Miller, C. 1987. "How Many Pollock in Donut Hole?" *Alaska Fisherman's Journal,* October 1987, 54–56.

Milner, R. "Huxley's Bulldog: The Battles of E. Ray Lankester (1846–1929)." *Anatomical Record* 257, no. 3 (1999): 90–95.

Miyake, H., K. Itaya, H. Asami, H. Shimada, M. Watanobe, T. Mutoh, and T. Nakatani. "Present Condition of Walleye Pollock Spawning Ground Formation in the Sea of Japan off Western Hokkaido, Viewed from the Recent Conditon of the Egg Distributions." *Bulletin of the Japanese Society of Fisheries Oceanography* 72 (2008): 265–72.

Molyneaux, P. *The Doryman's Reflection.* New York: Thunder's Mouth Press, 2005.

Morrell, Virginia. "Can Science Keep Alaska's Bering Sea Pollock Fishery Healthy?" *Science* 326 (2009): 1340–41. http://www.sciencemag.org/cgi/content/full/326/5958/1340?sa_campaign=Email/sntw/4-December-2009/10.1126/science.326.5958.1340x.

Mueter, F. J., N. A. Bond, J. N. Ianelli, and A. B. Hollowed. "Expected Declines in Recruitment of Walleye Pollock (*Theragra Chalcogramma*) in the Eastern Bering Sea under Future Climate Change." *ICES Journal of Marine Science* 68, no. 6 (Jul 2011): 1284–96.

Mulligan, T. J., K. M. Bailey, and S. Hinckley. "The Occurrence of Larval and Juvenile Walleye Pollock, *Theragra Chalcogramma*, in the Eastern Bering Sea with Implications for Stock Structure." In *Proc. Int. Symp. Biol. Mgmt. Walleye Pollock, Alaska Sea Grant Report* (1989): 471–90.

Mullon, C., P. Freon, and P. Cury. "The Dynamics of Collapse in World Fisheries." *Fish and Fisheries* 6 (2005): 111–20.

Munson, Fred. March 15, 2011.

Murphy, Kim. "Bering Sea: Where Few Even Dare." *Kitsap Sun,* April 28, 2001.

Myers, R. A., and B. Worm. "Rapid Worldwide Depletion of Predatory Fish Communities." *Nature* 423 (2003): 280–83.

Ness, Kaare. October 28, 2011.

Nicholson, A. J. "The Balance of Animal Populations." *Journal of Animal Ecology* 2 (1933): 131–78.

Nilsen, Andre. "The Norwegian Parliament Awards Prize to Corrupt Guantanamo Operator." Oxford Council on Good Governance, 2006. Described in: Flato, Hedda. "Tough Out against Rokke." NA24, July 28, 2006. http://www.na24.no/arkiv/naeringsliv/article695497.ece.

"Nippon Suisan Kaisha, Limited." Funding Universe. Accessed August 14, 2012. http://www.fundinguniverse.com/company-histories/NIPPON-SUISAN -KAISHA-LIMITED-Company-History.html.

Nishimura, A. Email, January 15, 2011.

NOAA Fisheries. "Final Environmental Impact Statement for American Fisheries Act, Amendments 61/61/13/8." Published 2002. http://www.fakr.noaa.gov/ sustainablefisheries/afa/eis2002.pdf

Norman, J. R. *A History of Fishes*. New York: A. A. Wyn, 1948.

North Pacific Fisheries Management Council. "Celebrating 30 Years of Sustainable Fisheries." Anchorage: North Pacific Fisheries Management Council, 2006.

"Norway Developing Eco-Friendly Trawl Technology." MercoPress. Published April 7, 2008. http://www.mercopress.com/vernoticia.do?id=13083& formato=html.

Okada, K. "Biological Characteristics and Abundance of the Pelagic Pollock in the Aleutian Basin." Paper presented at the International North Pacific Groundfish Symposium, 1983. Japan: Far Seas Fisheries Research Laboratory, 1983.

Onstot, Laura. "King of Fish Sticks." *Seattle Weekly*, November 19, 2008.

"Overfishing of Pollock Risks Collapse of World's Largest Food Fishery, Endangers Sea Lions and Seal." Greenpeace. Published December 2, 2008. http://www .greenpeace.org/usa/en/media-center/news-releases/overfishing-of-pollock -risks-c/.

"Owners Starting Bering Sea Fund." *Fishing News International*, August 1991, 64.

Pacific Seafood Processors Association. *A Strategy of the Americanization of the Groundfish Fisheries of the Northeast Pacific: Summary Report*. Seattle: Pacific Seafood Processors Association, 1985.

Parker, D. "US and Soviets Seek Plug for Donut Hole." *Pacific Fishing*, October 1991, 37.

Pauly, D. "One Hundred Million Tonnes of Fish, and Fisheries Research." *Fisheries Research* 25 (1996): 25–38.

Pauly, D. "Aquacalypse Now." *New Republic*, September 28, 2009. http://www.tnr .com/article/environment-energy/aquacalypse-now.

Pauly, D., V. Christensen, J. Dalsgaard, R. Froese, and F. C. Torres. "Fishing Down Marine Food Webs." *Science* 279 (1998): 860–63.

Pereyra, Walter. January 9, January 12, April 11, 2011.

PEW Oceans Commission. "A Dialogue on America's Fisheries." Arlington, Virginia: PEW Oceans Commission, 2003.

Plesha, Joseph. July 25, 2011.

"Pollock Cutbacks Sought—Environmentalists Want to Protect Sea Lions." *Seattle Times*, August 10, 1999.

"Pollock Fleet Boom Brings Big Business." *Fishing News International*, December 1988, 7–10.

"Raising a Ruckus!" Living on Earth. Published June 5, 1998. http://www.loe.org/ shows/segments.html?programID=98-P13-00023&segmentID=5.

Ramsey, Bruce. "Shore Fishing Plants vs. Trawlers—The Battle Gets Nasty." *Seattle Post-Intelligencer*, October 1, 1997.

Ramsey, Bruce. "The Tide of Change: Bering Sea Spawns New Fishing Era." *Seattle Post-Intelligencer*, May 2, 1988.

"Reports of Fisheries' Demise Are Greatly Exaggerated." At-Sea Processors Association. Published 2006. www.atsea.org/learnmore.php.

"Review of the State of World Fishery Resources: Marine Fisheries. Northwest Pacific." *FAO Fisheries Circular No. 920 FIRM/C920*, 1997.

Ricker, W. E. "Stock and Recruitment." *Journal of the Fisheries Research Board of Canada* 11 (1954): 559–623.

"Rise of American Seafoods Company." *Fishing News International*, October 1991, 16–21.

"Roekke and RGI." Greenpeace. Accessed August 3, 2012. http://archive .greenpeace.org/comms/fish/am03.html.

Rozwadowski, H. M. *The Sea Knows No Boundaries*. Copenhagen: International Council for the Exploration of the Sea, 2002.

Russell, F. S., A. J. Southward, G. T. Boalch, and E. I. Butler. "Changes in Biological Conditions in the English Channel off Plymouth during the Last Half Century." *Nature* 234 (1971): 468–70.

Ryther, J. H. "Photosynthesis and Fish Production in the Sea." *Science* 166 (1969): 72–76.

Sabella, John. "Taking the First Step: On the Road to Americanization." *Pacific Fishing*, October 1985, 43–49.

Saporito, Bill. "The Most Dangerous Job in America." In *CNN Money*. Published May 31, 1993. http://money.cnn.com/magazines/fortune/fortune_archive/ 1993/05/31/77905/index.htm.

Sasaki, T., and T. Yoshimura. "Past Progress and Present Condition of the Japanese Pollock Fishery in the Aleutian Basin." (Document submitted to the Annual Meeting of the International North Pacific Fisheries Commission, Vancouver, Canada, October 1987. Tokyo: Fisheries Agency of Japan, 1987.

Schaefer, David, and Duff Wilson. "Fish Quota Favors Alaska over Seattle Ships— Critics Call Ruling 'Blatant Politics.'" *Seattle Times*, March 5, 1992.

Schaefer, M. B. "Some Aspects of the Dynamics of Populations Important to Management of the Commercial Marine Fisheries." *Bulletin of the Inter-American Tropical Tuna Commission* 1, no. 2 (1954): 27–56.

Schindler, D. E., R. Hilborn, B. Chasco, et al. "Population Diversity and the Portfolio Effect in an Exploited Species." *Nature* 465 (2010): 609–12.

Schrank, W. E. The Newfoundland Fishery: Ten Years after the Moratorium. *Marine Policy* 29 (2005): 407–20.

Schwach, V., and J. M. Hubbard. "Johan Hjort and the Birth of Fisheries Biology: The Construction and Transfer of Knowledge, Approaches, and Attitudes, Norway and Canada, 1890–1920." *Studia Atlantica* 13 (2009): 22–41.

Schwach, Vera. March 6, 2011.

Sealy, T. S. "Soviet Fisheries: A Review." *Marine Fisheries Review* 36 (1974): 5–33.

"Seattle Will Make Its Crab Meat." *Fishing News International*, May 1985, 13.

"Senator Seeks Limits on Fleet of Huge Bering Sea Trawlers." *Seattle Times*, December 15, 1997.

Shafer, Sheila. "Barry Fisher: Still Outspoken after All These Years." *Pacific Fishing*, March 1986, 31–39.

Sharp, R., and U. R. Sumaila. "Quantification of US Marine Fisheries Subsidies." *North American Journal of Fisheries Management* 29, no. 1 (Feb 2009): 18–32.

Sims, Paul. "260M Servings of Supermarket Fish 'Could Be Wrongly Labelled.'" *Daily Mail*, April 25, 2011.

Sjong, John. January 28, 2011.

"Skipper's to Buy Alaska Pollock from Royal Sea." *Pacific Fishing*, November 1986, 21.

Skjæraasen, Jon Egil, Justin J. Meager, Ørjan Karlsen, Jeffrey A. Hutchings, and Anders Ferno. "Extreme Spawning-Site Fidelity in Atlantic Cod." *ICES Journal of Marine Science* 68, no. 7 ( 2011 ): 1472–77.

Smedbol, K.R., and J. S. Wroblewski. "Metapopulation Theory and Northern Cod Population Structure: Interdependency of Subpopulations in Recovery of a Groundfish Population." *Fisheries Research* 55 (2002): 161–74.

Smith, Roger Dale. "Navigating from Harbored to Heavy Seas: A History of Japan's International Fisheries in the North Pacific, 1900–1976." Master's thesis, University of British Columbia, 1999.

Smith, T. D. *Scaling Fisheries: The Science of Measuring the Effects of Fishing, 1855–1955*. Cambridge: Cambridge University Press, 1994.

Sogard, S. M., and B. L. Olla. "Effects of Light, Thermoclines, and Predator Presence on Vertical Distribution and Behavioral Interactions of Juvenile Walleye Pollock, *Theragra Chalcogramma* Pallas." *Journal of Experimental Marine Biology and Ecology* 167 (1993): 179–95.

Sogard, S. M., and B. L. Olla. "Food Deprivation Affects Vertical Distribution and Activity of a Marine Fish in a Thermal Gradient: Potential Energy-Conserving Mechanisms." *Marine Ecology Progress Series* 133 (1996): 43–55.

Sokolova, Alyona. "Seattle Fish Exec to Russians: 'Your Problem Is Corruption.'" *Pacific Fishing*, October 2006, 1.

Sonu, S. C. "Surimi." *NOAA Technical Memorandum NMFS*, January 1986.

Speiss, Ben. "Peace in the Pollock Business?" *Pacific Fishing*, November 1998, 30–32.

Speiss, Ben. "Triumph to Trump Critics." *Pacific Fishing*, August 1998, 32.

Springer, A. M. "A Review: Walleye Pollock in the North Pacific: How Much Difference Do They Really Make?" *Fisheries Oceanography* 5 (1992): 205–23.

Stavrum, Gunnar. *Kjell Inge Røkke: En Uautorisert Biografi*. Oslo: Glydendal, 1997.

Stephan, P. E., and S. G. Levin. *Striking the Mother Lode in Science: The Importance of Age, Place, and Time*. New York: Oxford University Press, 1992.

Stevens, Mick. February 21, 2011.

Storr-Paulsen, M., K. Wieland, H. Hovgard, and H. Rätz. "Stock Structure of

Atlantic Cod (*Gadus morhua*) in West Greenland Waters: Implications of Transport and Migration." *ICES Journal of Marine Science* 61 (2004): 972–82.

Stratton Commission. "Our Nation and the Sea." *Report of the Commission on Marine Science, Engineering, and Resources*, 1969.

Stuart, Sarah Clark. *Shell Game: How the Federal Government Is Hiding the Mismanagement of Our Nation's Fisheries*. Washington, DC: Marine Fish Conservation Network, 2006.

Stump, Ken. January 19, 2011.

Subcommittee on Fisheries Conservation, Wildlife, and Oceans. *Oversight Hearing on United States Ownership of Fishing Vessels*. Washington, DC: Government Printing Office, 1998.

Sudness, H. M. "From Bare Hands to Global Embrace." *World Fishing* 55, no. 2 (March 2006): 10–11.

Sunde, Scott. "Factory Trawler Fleet May Sit in Port Several Months." *Seattle Post-Intelligencer*, February 26, 1993.

"Surimi by Unisea Planned for Dutch Harbor." *Pacific Fishing*, August 1985, 16.

"Surimi Giant for US." *Fishing News International*, September 1986, 1.

Svedäng, H., M. Cardinale, and C. Andre. "Recovery of Former Fish Productivity: Philopatric Behaviors Put Depleted Stocks in an Unforseen Deadlock." In *Ecosystem-Based Management for Marine Fisheries: An Evolving Perspective*, edited by A. Belgrano and C. W. Fowler, 232–47. New York: Cambridge University Press, 2011.

"Taiyo Fishery Company, Limited." Funding Universe. Accessed August 7, 2012. http://www.fundinguniverse.com/company-histories/TAIYO-FISHERY -COMPANY-LIMITED-Company-History.html.

"A Tale of Two Fisheries." *Economist*, September 10, 2009. http://www.economist .com/sciencetechnology/PrinterFriendly.cfm?story_id=14401157 9/14/2009.

"Talks Heat Up on Illegal Japanese Fishing." *Pacific Fishing*, November 1988, 19.

Thompson, W. F. *The Effects of Fishing on Stocks of Halibut in the Pacific*. Seattle: Fisheries Research Institute, University of Washington, 1950.

Thompson, W. F., and F. H. Bell. "Biological Statistics of the Pacific Halibut Fishery, 2: Effect of Changes in Intensity upon Total Yield and Yield per Unit of Gear." *Report of the International Fisheries Commission* 8 (1934): 1–49.

"Tillion Leaves Council." *Pacific Fishing*, August 1997, 15.

Tollefsen, Astrid. *Following the Waters: Voices from the Final Norwegian Emigration*. Brewster, MA: Leifur Publications, 2005.

Tollesen, Tor. March 23 and April 23, 2011.

"Too Little, Too Late for Largest US Fishery." Greenpeace. Published October 1, 2009. http://www.greenpeace.org/usa/en/news-and-blogs/news/too-little -too-late-for-large/.

Toth, R. C. "Attack May Have Caught Kremlin by Surprise." *Seattle Times*, September 2, 1983.

Touza, Ann. "Fishing on the Far Side of the World." *Pacific Fishing*, April 2011, 9–12.

"Trawlers: Allocation of Pollock Ruining Us." *Seattle Times*, November 11, 1994.

Traynor, J. J., W. A. Karp, T. M. Sample, et al. "Methodology and Biological Results from Surveys of Walleye Pollock (*Theragra Chalcogramma*) in the Eastern Bering Sea and Aleutian Basin in 1988." Paper presented at the Proceedings of the Symposium on Application of Stock Assessment Techniques to Gadids, Seattle, Washington, 1989.

Uchida, H., and M. Watanabe. "Walleye Pollack [*sic*](Suketoudara) Fishery Management in the Hiyama Region of Hokkaido, Japan." In *Case Studies in Fisheries Self-Governance*, edited by R. Townsend, R. Shotton, and H. Uchida. FAO Fisheries Technical Paper, 163–74. Rome: FAO, 2008.

Urch, Mike. "Time to Be Upbeat about Fisheries." SeafoodSource.com. Published November 7, 2011. http://www.seafoodsource.com/newsarticledetail.aspx?id =12841.

US Census Bureau. "State and County Quick Facts." 2010. http://quickfacts .census.gov/qfd/index.html.

"US Surimi Plant in Production." *Fishing News International*, June 1986, 1.

Vaisman, A. "Trawling in the Mist: Industrial Fisheries in the Russian Part of the Bering Sea." TRAFFIC Network Report. Published 2001. http://www.traffic .org/fisheries-reports/traffic_pub_fisheries5.pdf.

Volterra, V. "Fluctuations in the Abundance of a Species Considered Mathematically." *Nature* 118 (1926): 558–60.

Von Munchow, Otto. "Norwegian American: Kaare Ness." Published 2007. Norway.Com.

Warren, Brad. "How the Council Deals Out Pollock Options." *Pacific Fishing*, November 1997, 33.

Warren, Brad. "New Twist on Groundfish Wars." *National Fisherman*, August 1992, 11–13.

Warrenchuk, John. December 17, 2010.

Watling, L., and E. Norse. "Disturbance of the Seabed by Mobile Fishing Gear: A Comparison to Forest Clearcutting." *Conservation Biology* 12 (1998): 1180–97.

"We're Out at the Stroke of a Pen." *Fishing News International*, October 1991, 9.

Wespestad, V. G. "The Status of Bering Sea Pollock and the Effect of the 'Donut Hole' Fishery." *Fisheries* 18(3) 1993: 18–24.

Weber, M. *From Abundance to Scarcity*. Washington, DC: Island Press, 2001.

Western Alaska Community Development Association. *Western Alaska Community Development Quota Program*. Anchorage: Western Alaska Community Development Association, 2010.

Westneat, Danny. "Greenpeace Blockades Trawlers—Activists Chain Propellers to Halt Fishing Trips." *Seattle Times*, August 17, 1996.

Whitney, David. "Battle for the Bering Sea—Stevens Says Bill Would Protect Fishery, American Interests." *Anchorage Daily News*, December 14, 1997.

World Wildlife Fund (WWF). "Illegal Fishers Plunder the Arctic." Published April 16, 2008. http://wwf.panda.org/wwf_news/press_releases/?131061.

Worm, B., E. B. Barbier, N. Beaumont, et al. "Impacts of Biodiversity Loss on Ocean Ecosystem Services." *Science* 314, no. 5800 (2006): 787–90.

Worm, B., R. Hilborn, J. K. Baum, et al. "Rebuilding Global Fisheries." *Science* 325, no. 5940 (2009): 578–85.

Wyman, Jebb. "Fracas over US–Russian Boundary." *Pacific Fishing*, November 1999, 26.

Yatsu, A. "Japan." In *Impacts of Climate and Climate Change on Key Species*, edited by R.J. Beamish, 57–71. Vancouver, BC: PICES Press, 2008.

Zhang, C. I., and S. Kim. "A Pragmatic Approach for Ecosystem-Based Fisheries Assessment and Management: A Korean Marine Ranch Ecosystem." In *Ecosystem-Based Mangement for Marine Fisheries: An Evolving Perspective*, edited by A. Belgrano and C.W. Fowler, 153–80. New York: Cambridge University Press, 2011.

# Index